PRAISE FOR

Sepp Holzer's Permaculture

"A fascinating book written by a man who has devoted a lifetime to working with nature and creating extraordinarily diverse polycultures. His work is breathtaking."
—MADDY HARLAND,
editor of *Permaculture Magazine*

"At an altitude that everyone else has abandoned to low-value forestry, [Sepp Holzer's] is probably the best example of a permaculture farm in Europe [and] stands out like a beacon."
—PATRICK WHITEFIELD,
permaculture author and teacher

"As the era of cheap energy, stable climates, and surplus fertilizer stocks comes to a close, the principles of permaculture will become increasingly attractive as one way to design our future food and agriculture systems. This book provides important insights for applying these principles, for both rural farming and emerging forms of urban agriculture."
—FREDERICK KIRSCHENMANN, president of
Stone Barns Center for Food and Agriculture

"Sepp Holzer is a Superstar Farmer who turns out an absolutely remarkable volume and variety of food products without one smidgen of chemical fertilizer, and on land in Austria that an Illinois corn farmer would pronounce too marginal for agriculture."
—GENE LOGSDON, author of *Holy Shit: Managing Manure
to Save Mankind* and *The Contrary Farmer*

"Here's great news for fruit-loving gardeners everywhere! Most of the work of establishing, pruning, and tending fruit trees by 'modern' methods is unnecessary and even counterproductive. This book is *The One-Straw Revolution* for tree crops."
—CAROL DEPPE, author of *The Resilient Gardener:
Food Production and Self-Reliance in Uncertain Times*

Sepp Holzer's Permaculture

*A Practical Guide
to Small-Scale, Integrative Farming
and Gardening*

Chelsea Green Publishing
White River Junction, Vermont

First English language edition © 2010 by Permanent Publications. www.permaculture.co.uk

The right of Sepp Holzer to be identified as the author of this work has been asserted by him in accordance with
the Copyrights, Designs and Patents Act 1998.

The first English language edition of *Sepp Holzer's Permaculture* was published in 2011 in the United Kingdom by
Permanent Publications, The Sustainability Centre, East Meon, Hampshire GU32 1HR, UK.
www.permaculture.co.uk.

Translated from the German by Anna Sapsford-Francis

Indexer: Linda Hallinger

Designer: Two Plus George Limited, www.TwoPlusGeorge.co.uk

Printed in the United States of America
First Chelsea Green printing March, 2011
13 12 11 10   20 21 22 23

Our Commitment to Green Publishing
Chelsea Green sees publishing as a tool for cultural change and ecological stewardship.
We strive to align our book manufacturing practices with our editorial mission and to
reduce the impact of our business enterprise in the environment. We print our books and
catalogs on chlorine-free recycled paper, using vegetable-based inks whenever possible.
This book may cost slightly more because it was printed on paper containing recycled fi-
bers, and we hope you'll agree that it's worth it. *Sepp Holzer's Permaculture* was printed on
FSC®-certified paper supplied by LSC Communications that contains at least 10-percent
postconsumer recycled fiber.

Library of Congress Cataloging-in-Publication Data
Holzer, Sepp.
 [Sepp Holzers Permakultur. English]
 Sepp Holzer's permaculture : a practical guide to small-scale, integrative farming and gardening / [Sepp Holzer].
-- 1st English language ed.
    p. cm.
 First published in German as Sepp Holzers Permakultur by Leopold Stocker Verlag.
 ISBN 978-1-60358-370-1
1. Permaculture. I. Title. II. Title: Permaculture.

S494.5.P47H6813 2001
631.5'8--dc22

                                    2011006356

Chelsea Green Publishing
85 North Main Street, Suite 120
White River Junction, VT 05001
(802) 295-6300
www.chelseagreen.com

MIX
Paper from
responsible sources
FSC® C132124
www.fsc.org

Contents

2 Alternative Agriculture

3　Fruit Trees

4　Cultivating Mushrooms

5 Gardens

6 Projects

Foreword

In the film *Sepp Holzer's Permaculture* there's an image that really says it all. Taken from the air, it shows steep mountainsides covered in seemingly endless monocultures of spruce trees, broken only by the mountainside that is Sepp Holzer's farm, the Krameterhof. In contrast to the dark trees on either side, it's an intricate network of terraces, raised beds, ponds, waterways and tracks, well covered with fruit trees and other productive vegetation and with the farmhouse neatly nestling amongst them. There, at an altitude which everyone else has abandoned to low-value forestry, what is probably the best example of a permaculture farm in Europe stands out like a beacon. It stands as witness to both the contrariness and the skill of the Rebel Farmer.

He has always gone against the grain of modern farming: he cultivates rich mixtures of plants and animals in place of monocultures; he has no need for chemicals because the dynamic interactions between the plants and animals in his polycultures provide all the services which conventional farmers find in the fertiliser bag and the crop sprayer. But it takes more than a contrary nature to be a rebel farmer. It also takes skill and knowledge, and these don't come easily. Right from his childhood, when his mother gave him a small plot for his first garden, he has observed, questioned, experimented, observed again and experimented again. He knows the natural world like few other people do today, and treats his farm as an integrated part of that natural world – which is exactly what it is.

In this book he shares the skill and knowledge which he has acquired over his lifetime. He covers every aspect of his farming, not just how he creates a holistic system on the farm itself but also how he makes a living from it. He writes about everything from the overall concepts which guide him down to the details, such as which fruit varieties he has found best for permaculture growing. Farming at such a high altitude is a challenge in itself, and as well as his knowledge of plant and animal interactions he has had to learn much about how heat and water pass through the ecosystem, and how they can be stored and made to work for the system.

An important part of permaculture is getting to know your own individual place. Every patch of the Earth has its unique personality and character, just as each person has. Nevertheless Sepp Holzer has taken his skill and applied it on sites all over the world and in urban gardens too. It takes a great deal of experience to be able to look at a site in a different part of the world and understand how it works well enough to be able to give advice on it.

The other side of that coin is that what works for him on his Austrian mountain will not necessarily work for you on your own land. Here in Britain, for example, we have a cloudy maritime climate, in strong contrast to Austria's continental climate. Although our winters are milder, so too are our summers. Above all we lack the sunshine which is such a key element in the way he creates favourable microclimates. Humidity is also greater here. What you can do, say, at 250m on the edge of Bodmin Moor is not the same as what you can do at ten times that altitude on the Krameterhof. Similar allowances must be made for other parts of the world.

This is not to negate the value of this book for people who live outside Austria – far from it. Much of the detailed information is highly relevant in any temperate country. As long as you bear in mind that both your climate and your soil are possibly quite different to those on the Kameterhof, you will find it a storehouse of valuable information.

Nevertheless the book's greatest value is not so much in the information it contains but in the attitudes it teaches. Its message is not so much 'this is how you do it' but 'this is the way you go about thinking of how to do it.' Sepp Holzer's way is the way of the future. In the fossil fuel age we've been able to impose our will on the land by throwing cheap energy at every problem. In the future that option won't be open to us any more. We'll have to tread the more subtle path, the path which patiently observes nature and seeks to imitate it. That future may not be as far off as we think.

Patrick Whitefield
September 2010

Patrick Whitefield is a permaculture teacher and the author of *Permaculture in a Nutshell* (1993), *How to make a Forest Garden* (1996), *The Earth Care Manual* (2004) and *The Living Landscape* (2009). More details about his courses can be found at www.patrickwhitefield.co.uk

Preface

Dear readers,

This is the second book I have written so far, to pass on my over 40 years of experience as a farmer practising alternative agriculture. I was inspired to do this by the many people who have come to visit the Krameterhof: among them teachers, professors and doctors as well as farmers and gardeners. My darling wife, Vroni, and my children were particularly eager for me to put my experiences and discoveries into writing. My first book, an autobiography entitled *The Rebel Farmer*, sold over 120,000 copies in just under two years and was a great success. It was presented with a golden book award and I received well over a thousand letters from enthusiastic readers. This made me realise that there was a great deal of interest in my work. When my daughter Claudia and son Josef Andreas offered their help, I simply could not refuse.

Sepp Holzer

I want this book to help people realise that trying to understand and live in harmony with nature instead of fighting against it is well worth the effort. On the countless trips I have made to oversee my projects abroad, I have seen many terrible sights which have stayed with me and even given me a few nightmares. Whether it is in Bosnia, Colombia, Brazil, Thailand or in the United States, it is plain to see how irresponsibly nature is treated everywhere. Many people seem to have lost their ability to think independently about or to feel responsibility for our planet and its future. The result is a loss of respect for nature and our fellow creatures. Tens of thousands of hectares of scrubland and rainforest are intentionally being burned to make way for monocultures and, of course, any wildlife is destroyed along with it. A small few profit at the expense of large swathes of society, who generally do not know how to provide themselves with food. The poverty and hardship people endure in 'developing countries' knows no bounds! Young and old alike are treated like refuse and live on the streets from hand to mouth. Only the powerful have rights, which I have seen and experienced for myself. In addition, this happens in areas where no one should have to go hungry, because the soil is fertile and the weather is favourable. There should be more than enough food for everyone. Many people have lost their land to powerful landowners and with that the ability to provide for their families. They have had their independence taken away, which then becomes very difficult to take back. So many of these people live rough on the outskirts of town in terrible conditions, whilst their land is relentlessly overworked and ruined.

Many people think that this cannot happen in Europe, but we are already well on our way! Most small farms only provide a subsidiary income, because the farmers do not know how to make enough money from them to live on any more. Today, very few people dare to forge their own way and consider alternative farming methods. Instead many people look to subsidy programmes to tell them how to run their businesses and alter their farms accordingly. Either that or quantity is prioritised over quality and farmers try to compensate for low prices with a larger volume of produce. The result is a monoculture maintained with large quantities of chemicals. Many people are deterred by the bureaucratic obstacles that are put in their way when they try to practice alternative farming methods. It is every person's duty to defend their rights, land and even their concept of democracy and make them their guiding principles. If we do not, there is a real danger of finding ourselves in an administrative and bureaucratic dictatorship.

I have already described how difficult it is to forge your own way in my first book. Some years ago I had a visitor from New Zealand. This visitor was the late Joe Polaischer – our lives took similar paths. He chose to leave Austria and emigrate to New Zealand to set up a permaculture farm under difficult conditions. He had visitors all the way from Europe and they were delighted with what he had accomplished. Joe was a remarkable man. He was a teacher and had a great deal of practical experience, which is exactly what we need right now. His achievements should make it clear that there are people on the other side of the world who want to live in harmony with their environment and not at odds with it. Treating our planet and fellow creatures with respect – and not being motivated by rivalry, jealousy or hatred – is the only way!

My dear friend Joe, for your commitment to using land sustainably, your contribution to the development and teaching of permaculture in Austria, you have my most heartfelt thanks.

I would also like to thank my colleagues of many years Erich Auernig and Elisabeth Mohr, who have always supported me in my work. Without their tireless efforts it would never have been possible to raise such a large amount of public interest in my farming methods. With their help, I have been able to show thousands of interested visitors around the Krameterhof and oversee countless projects abroad. I have also had the opportunity to pass on my experience through presentations and seminars. I would also like to express my gratitude to Mrs Maria Kendlbacher and her daughter Heidi who look after our guests on the Krameterhof. I also thank my brother and gamekeeper Martin Holzer.

Most of all I would like to thank my family and my darling Vroni! Throughout our more than 36 years of marriage she has always stood by me and supported me completely. Without her it would never have been possible to run the Krameterhof so successfully and still have time to write a book. It is a joy to have such a wonderful family.

In this book I have tried to answer the most frequent questions raised at my presentations and seminars. I hope that this book helps you to find your way towards living a life in harmony with nature: whether it begins with a windowbox, a garden, or a field is not important. If this book helps one person to start thinking ecologically and independently, it will have done its job. I wish you success putting your ideas and, perhaps, permaculture projects of your own into practice.

General Conversion Formulae

From	To	Multiply by
inches	millimetres	25.4
millimetres	inches	0.0394
inches	centimetres	2.54
centimetres	inches	0.3937
feet	metres	0.3048
metres	feet	3.281
yards	metres	0.9144
metres	yards	1.094
sq inches	sq centimetres	6.452
sq centimetres	sq inches	0.155
sq metres	sq feet	10.76
sq feet	sq metres	0.0929
sq yards	sq metres	0.8361
sq metres	sq yards	1.196
acres	hectares	0.4047
hectares	acres	2.471
pints	litres	0.5682
litres	pints	1.76
gallons	litres	4.546
litres	gallons	0.22
ounces	grams	28.35
grams	ounces	0.03527
pounds	grams	453.6
grams	pounds	0.002205
pounds	kilograms	0.4536
kilograms	pounds	2.205

Introduction

In 1962, at the age of 19, I took over my parents' farm in Lungau, Salzburg. Since then I have managed the Krameterhof in my own way. I have built ponds, terraces and gardens, kept fish and wild cattle, I have grown mushrooms, set up an alternative tree nursery and so much more. Despite the fact that there are many different areas a farm can specialise in, it was important to me that I did not focus on any one source of income. I wanted to remain as flexible as possible, so that I would always be able to react to changing market conditions. In addition, my interests at the time were so broad that there was no way I would have been able to decide on just one area. Over the years, this decision has been proved right again and again. It is true that many people called me 'crazy' during my time as a young farmer. They said that my methods would not amount to much and that I would soon have to sell the farm, but success proved me right in the end. Since then I have managed to double the original size of the Krameterhof, whilst many of my critics have had to give up their farms or look for additional income. Now the Krameterhof measures 45 hectares, reaching from 1,000 to 1,500m above sea level across the southern slope of the Schwarzenberg mountain. People still call me 'crazy' today, but it does not really bother me any more. I have realised that many people find it difficult to accept when you do things in a way that is not so widely recognised. This makes you difficult to predict and harder to control, which many people find threatening.

My alternative farming methods have brought me into conflict with the authorities many times and some of these disputes have been extremely drawn-out and tiring. It has taken a great deal of strength and effort to come through them and to not let myself be discouraged. One conflict with our self-important administrative system, which was making my life as an independent farmer difficult, caused me many sleepless nights. Times were often difficult and I did not know how I was going to get through it all. Fortunately, my wife Veronika always supported me completely and has stood by me all of these years, which has given me the strength to carry on despite the conditions set by the authorities, the special taxes and other chicanery. I also gathered strength from nature: whenever I had finished with yet another tedious lawsuit or had read one of the many impracticable expert reports, I would wander through my cultures and, for hours, collect seeds and sow them again in different places. Observing my plants and livestock also gave me fresh energy. Nature and my family have helped me to persevere despite the nightmarish bureaucracy. It is incomprehensible to me that a person with so many innovative ideas should have so many hurdles and stumbling blocks put in their way. The fact that I have not let myself be intimidated and do not stay quiet just to please people has given me a reputation for being a 'rebel farmer'. The fact that it is actually necessary to become a 'rebel' to run a farm in harmony with nature is really very sad! The administrative system has become overgrown and nips any creative thought in the bud. It is the responsibility of those in power to fix these problems.

We have to make democracy our guiding principle instead of acting like lemmings and following the masses blindly, otherwise one day we will lose our democracy and our rights. On my farm I have no problems with large populations of 'pests', because nature is perfect and keeps everything in balance. I only wish that our administrative system could be regulated in a similar way, so that the bureaucracy does not push us to breaking point and we are not punished for thinking creatively. I think we all need to work to combat this unbearable situation and bring this 'bureaucratic overpopulation' back under normal levels.

In the summer of 1995, I received a letter from the University of Natural Resources and Applied Life Sciences in Vienna asking if they could hold a seminar at the Krameterhof. Through this seminar I learned for the first time that there was a term for my farming methods: 'permaculture'. This word was coined by the Australian ecologist Bill Mollison and his student David Holmgren and is derived from 'permanent agriculture'. A permaculture system is a system that resembles nature and is based on natural cycles and ecosystems. Some of the students from the seminar sent me a few books on permaculture. As I read the books I could only agree with the arguments within them. The fundamental thoughts and ideas in these books were incredibly similar to my own methods. I discovered that whilst there are many new farms, which claim to use 'permaculture' methods, there was not a single one that worked in the same way as ours on the Krameterhof. This is because the concept of permaculture was first developed in 1978, whereas I began to create gardens and ponds and experiment with sustainable systems in my youth. My methods have had over 40 years to develop. I have had time to continually improve upon and develop them

Veronika and Sepp Holzer

so that now I have as little work to do as possible and I still achieve good yields. It was obvious to me that I was doing this by imitating natural cycles. What aspect of nature could I improve upon when nature already functions perfectly? Every time I tried to improve upon nature I quickly realised that I had only created more work for myself and the loss in yield was greater. So I always returned to the natural way, which, as far as I am concerned, has proved to be the only right one.

- The basic principles of permaculture are:
- All of the elements within a system interact with each other.
- Multifunctionality: every element fulfils multiple functions and every function is performed by multiple elements.
- Use energy practically and efficiently, work with renewable energy.
- Use natural resources.
- Intensive systems in a small area.
- Utilise and shape natural processes and cycles.
- Support and use edge effects (creating highly productive small-scale structures).
- Diversity instead of monoculture.

My farming methods meet all of these criteria. When it was finally suggested that I should label my farm as being based on permaculture principles and open it to the public, I agreed.

Unfortunately, I soon found out that there are many self-styled permaculturists and permaculture designers who only concern themselves with permaculture theory and have no idea how to put it into practice. In permaculture design practical experience is indispensable. It is difficult to gain an understanding of nature just from theories. Only those with personal experience can give a professional consultation. So I think it is only appropriate for someone to offer their services as a permaculture designer if they have gathered plenty of practical experience over a number of years. A little work experience and a few short courses are certainly not enough. This is why I advise anyone interested in permaculture principles to find out how much practical experience these consultants have and not just rely on testimonials or other references. It is a good idea to take a look at the consultant or designer's offices in person before the consultation. This will tell you a great deal about their knowledge and abilities.

Holzer permaculture incorporates landscape design (creating terraces, raised beds, water gardens, ponds, humus storage ditches and microclimates), agroforestry (integrating trees and shrubs into farming), fishery, growing aquatic plants, keeping livestock, fruit-growing, alpine pastures and growing alpine and medicinal plants. Even tourism is not ruled out. Economy and ecology are not a contradiction. Holzer permaculture dates back to 1962 and is based on decades of experience running a full-time farm. You must see and understand this technique as a whole, so that it can be used profitably. Only those who practice permaculture can also understand it and pass it on to others. This is why it makes no sense to simply create a permaculture system just like mine. You must learn it for yourself like learning the alphabet at school. This is the only way you can achieve success and gain happiness from it. Permaculture principles work all over the world, as I have seen whilst working on my projects in Colombia, Thailand, Brazil, the United States and Scotland.

You can find up to date information on my projects as well as lectures, seminars and guided walks around the Krameterhof on our website www.krameterhof.at/en. Unfortunately, as a result of the large amount of public interest we can no longer answer all of the letters and inquiries that reach us. We ask for your understanding and hope that this book can answer at least some of these questions.

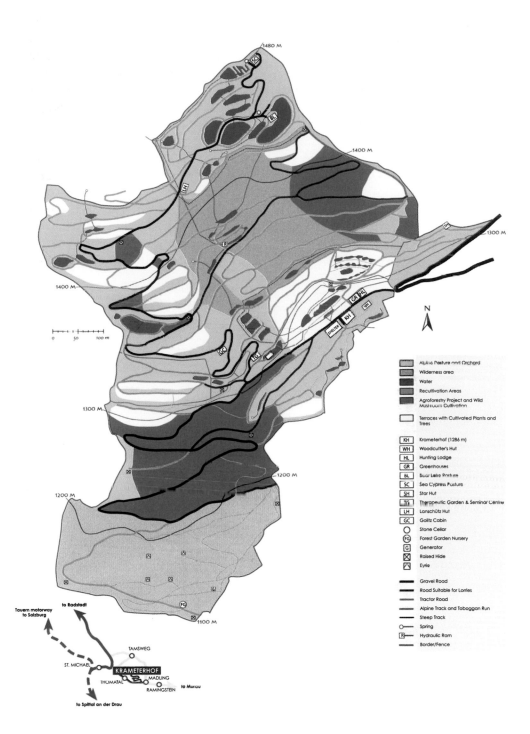

Alpine Pasture and Orchard

Wilderness area

Water

Recultivation Areas

Agroforestry Project and Wild Mushroom Cultivation

Terraces with Cultivated Plants and Trees

KH	Krameterhof (1286 m)
WH	Woodcutter's Hut
HL	Hunting Lodge
GR	Greenhouses
BL	Bear Lake Pasture
SC	Sea Cypress Pasture
SH	Star Hut
Th	Therapeutic Garden & Seminar Centre
LH	Lanschütz Hut
GC	Golitz Cabin
O	Stone Cellar
FG	Forest Garden Nursery
G	Generator
⊠	Raised Hide
	Eyrie

Gravel Road

Road Suitable for Lorries

Tractor Road

Alpine Track and Toboggan Run

Steep Track

Spring

Hydraulic Ram

Border/Fence

Tauern motorway to Salzburg

to Radstadt

TAMSWEG

ST. MICHAEL

KRAMETERHOF

THOMATAL   MADLING

RAMINGSTEIN   to Murau

to Spittal an der Drau

1 Landscape Design

Early Childhood Experiences

My first experiences with growing plants date back to childhood. I had a small piece of dry and stony land on a steep slope near the house that I rented from my father for two Austrian schillings, which would be a little over two pounds today. As far as my parents were concerned, the land was nigh on worthless and it was only cut back once a year. The sunny and stony plot provided an ideal habitat for a large population of snakes and it was for this reason that it was referred to as the *Beißwurmboanling* after the legendary *Beißwurm*, a large and poisonous serpent. As a child, this plot of land allowed me to learn more about

A photo of the Krameterhof from my childhood.

growing plants. Before this, my only experiences with cultivation came from tending my mother's flowerpots. I began to work my 2m² piece of land with a hoe and mattock. I laid down stones to make a bed, which, though small, was the first terrace I ever created. Strawberries, small fruit trees, pumpkins and many other plants soon began to grow. I noticed that the strawberries that grew very close to the stones were larger and sweeter than they would be usually. I named the strawberries 'stone strawberries' or *Stoaroadbe* and I traded them at school for erasers and Karl May books.

My experience with the strawberries made me realise something important which would be of use to me later on. A number of factors such as the stones' ability to store and release heat, their balancing effect on the temperature, the way the earth beneath the stones would remain wet and the abundance of earthworms and soil life all had a positive effect on the strawberries and would also have the same effect on other plants. This meant that my woodland strawberries, which would normally bear only small fruit, grew large and very sweet. This is why it is important to always observe the soil and plants closely. You should try whenever you can to find out why plants grow well, as well as discovering why plants grow poorly. This knowledge will help you to draw the

The Krameterhof in winter.

right conclusions. This is one of the most important skills needed for working with nature. It is still worth analysing things that have gone badly, because from this we can work out why they have failed. Why is this plant so beautiful and healthy while the other one is so sickly and weak? Why is one plant so lush and its leaves such a dark green, but the other so colourless and pale? Keeping a keen eye on the garden was a decisive factor in my success from the beginning. The range of plants I grew constantly increased and soon I was able to grow many different fruit trees, herbs and vegetables. I continued to observe and improve upon my system. Finally, I made my first pond so that I could breed my own fish. I have already written in detail about the way I developed my cultivation and landscaping methods throughout my childhood and youth in my first book, *The Rebel Farmer*.

Past Mistakes

In the last few decades many mistakes have been made with the management of land. In the name of agriculture we try to correct perceived imperfections in the terrain and drain unwanted water. Rocks and cliffs are blasted to make the fields and meadows suitable for mechanised farming. Wetlands in which the most beautiful orchids grow are drained and dense spruce monocultures are planted there instead. The Austrian Chamber of Agriculture is responsible for these measures having increased from 60% to 80% today. Large-scale drainage is still actively encouraged in a number of places. Hedges and orchards are still being grubbed out and cut down, rivers and streams are straightened and it is the monoculture system that is driving these changes.

The result of these monocultures and this irresponsible attitude to nature is already well known: the catastrophes are becoming greater and greater and the damage to the national economy is immeasurable. Floods, landslides and damage caused by storms and snow are becoming more prevalent. Valuable biomass and fertile humus are being lost. This narrow-minded attitude is causing the soil to lose its capacity to store water – entire areas of land are acidifying and turning to desert. Eventually, the widespread use of pesticides and fertilisers will poison the ground water. Biodiversity is being seriously threatened in these areas: instead of well-structured habitats there is suddenly only a monocultural landscape. This loss of habitat causes the population of a few plants and animals to increase rapidly whilst others disappear completely. Animal and plant diversity is being lost. Humans disturb the balance of nature and then begin to fight against 'pests' and 'weeds' for which they have only themselves to blame. A new industry, agrochemicals, has devoted itself to the destruction of these enemies by chemical means. Anyone who understands the processes of nature must recognise that it is we humans who have caused these organisms to appear in such large numbers. If conditions are conducive to only a few species then they are the ones that will become the most pervasive. Their natural predators and rivals, which help to keep the system in balance, are gone.

How can we even begin to rectify these mistakes? Recognising and admitting our mistakes is a step in the right direction. Once we have realised how far we have gone wrong, we can find our way back to natural ways of thinking and behaviour. It helps to focus on our own mistakes and not those of others. Even I have made many mistakes in the name of so-called 'modern agriculture'. I learnt about the assumptions modern agriculture has made from courses at

The terraces on the Krameterhof stretch from the valley (1,100m above sea level) to the mountain pasture (1,500m above sea level). Previously inaccessible areas can now be reached; this way all of the land can be used effectively.

The landscape was designed so that it would work in harmony with nature: the picture shows a wetland I created on the Krameterhof at 1,400m above sea level.

agricultural college, through training and from textbooks. As a young farmer, the Chamber of Agriculture used their biased subsidy system and various economic advisors to urge me to use modern cultivation techniques. I was encouraged to be a modern farmer and not a 'country bumpkin'. I let myself be convinced by one-sided information and biased instructors and I lost my way for some time.

Luckily, my experience with plants and animals began at a young age. This experience made me realise that I was on the wrong path. The damage I had caused before realising this was still limited. Nevertheless, if I had not had my own positive experiences to fall back on, I would still be on the path of so-called 'progress' whilst being completely unaware of the consequences.

I threw out the official guidelines and decided to restore the farm according to my own ideas. It was important to me that I had healthy and hardy plants and animals on the farm again. At last I could put my ideas into practice without answering to anyone else and just enjoy the process of cultivation. I started by using machines to improve and widen the terraces I made as a child. Then I wanted to make proper use of the springs on my land. I have always liked keeping fish, so I made ponds and lakes across the farm to breed them in. This is how the permaculture landscape of the Krameterhof began to emerge.

At the time I did not know about the rice terraces of Asia or the terraced fields of the Berbers in Morocco. It was only later that I discovered that this had been a tried and tested method of cultivation for millennia. I am convinced that anyone who tries to farm seriously will automatically arrive at the same methods. Many different cultures throughout the world have developed these successful and harmonious systems through trial and error and have constantly improved upon them by learning from their own mistakes.

Permaculture landscape design essentially involves restoring a partially destroyed natural landscape. It is about returning to small-scale landscapes based on natural ecosystems. It offers us a viable alternative to the monoculture system that destroys our soil and pollutes our ground water.

The Permaculture Landscape

General

A permaculture landscape is designed so that all of the plants and animals living there will work in harmony with each other. This is the only way to manage land in a stable and sustainable way. All available resources – whether they are springs, ponds, marshes, cliffs, forests or buildings – are used and included in the plan. It is important that the resources are used in a way that is appropriate to the environment around them; in practice this means that the natural features of each area must be supported and reinforced. To make proper use of the available natural resources we have to work with nature and not against it. This gives us the desired result for the least expenditure of energy.

Water is life and must therefore be treated with great care. This is why I try to keep water (whether it is rain water, spring water or surface runoff) on my land for as long as possible. There are many possible uses for this water. Where there is wet soil, for instance, I would make a pond, water garden or wetland and plant orchids. In dry places I grow herbs that prefer semi-arid conditions like thyme, creeping thyme (*Thymus serpyllum*), marjoram and sage. Grain amaranth and New Zealand spinach are also suited to dry places and give a good yield. These are just a few examples of plants that thrive in these conditions.

Terraces are a very important part of my permaculture system. Without terraces it would have been impossible for me to work the otherwise unproductive and at times inaccessible land on the Krameterhof. With these terraces, which can also be used as paths, I can cultivate even the steepest of slopes and still make a profit. The terraces even make it possible to use medium-sized machinery. They provide me with a substantially larger area for cultivation and gaining this extra land is particularly important for small farms. The terraces also help to stop valuable humus from being washed away or otherwise lost. Finally, they help to prevent soil erosion and make a considerable contribution to the health and fertility of the soil. When making a terrace it is very important to minimise the number of dead ends. If possible every terrace should form a continuous belt of land, so that the terraces can be worked using the least amount of energy. Whilst making terraces I try to follow the principles of nature. As a rule, there should be no straight lines, corners or steep slopes (with the exception of raised beds). It is also important to break up the landscape by creating plenty

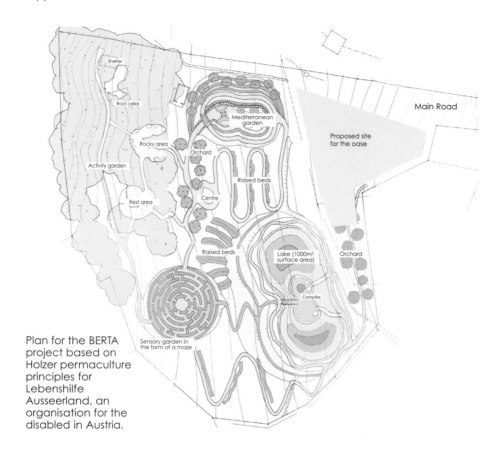

Plan for the BERTA project based on Holzer permaculture principles for Lebenshilfe Ausseerland, an organisation for the disabled in Austria.

Labels in image: Shelter, Root cellar, Mediterranean garden, Rocky area, Orchard, Activity garden, Raised beds, Rest area, Centre, Raised beds, Lake (1000m² surface area), Orchard, Campfire, Sensory garden in the form of a maze, Main Road, Proposed site for the oase

of forms and features. These help to create numerous microclimates, which give the land an even greater potential for cultivation. Creating dry areas, wetlands, hedges, windbreaks or raised beds in different locations results in special climatic conditions. In these places I can grow a large variety of plants that would otherwise not be able to survive.

The landscaping possibilities are almost limitless when creating a permaculture system. Anything is possible as long as the terrain and soil conditions allow it. Raised beds are used to grow vegetables and crops. Terraces provide a larger area for planting and access to the remotest corners of my farm. The beds and terraces can perform a multitude of different functions. For example, if a road or railway line borders the land, or if there is a factory nearby, I can use raised beds to keep out emissions, dust, noise and fumes. I place the beds on the edge of my land and plant them with various trees and shrubs. The beds and their lush vegetation work as a visual barrier and protects the land from pollution. They grow into a hedge that provides birds, hedgehogs and insects with shelter and somewhere to live. Barriers like this play a substantial part in encouraging communities of useful animals and insects.

Possible ways
to use the land
at Holzer Hof
in Burgenland,
Austria.

The permaculture system at Holzer Hof

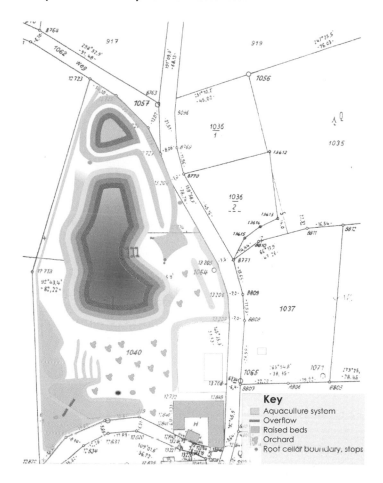

Key
- Aquaculture system
- Overflow
- Raised beds
- Orchard
- Root cellar boundary, stops

When designing beds and terraces you should respect property boundaries and your neighbours' rights. This way you will not have to deal with any unnecessary problems later on. I try to follow regulations as much as I can. If this is not done at the planning stage or whilst building, it will be a lot more difficult to gain the necessary permission from the authorities afterwards.

Livestock (pigs, chickens, ducks etc.) play an integral role in a permaculture system and they have had this importance for many cultures throughout history. A well thought-out system of paddocks and enclosures allows me to keep animals on the same land as my crops. Instead of damaging the land by overgrazing it, the animals actually help me with my work.

Naturally, great care must be taken when making larger changes to the landscape. The geological conditions must be taken into consideration to avoid landslides or gully erosion. As a result, it is always a good idea to start off slowly and gradually gather the experience you need. If you want to do something on a larger scale straight away you should seek professional advice. To this day, I have never seen a piece of land where it was not possible to establish and maintain a permaculture system. Whether it is in the smallest of gardens or on the largest of estates, in town or in the country, permaculture principles can be applied anywhere.

Reshaping the Land with an Excavator

According to Holzer permaculture principles, mechanical diggers only need to be used once when the system is being set up. The depth you can dig to depends on local soil conditions. This can vary from 20cm to two or three metres depending on what kind of structure you are building. The legal requirements for this sort of work vary from state to state in Austria. These requirements seem largely pointless to me, because they can be interpreted in many different ways and most of them are unrealistic. If you have a project that goes against any of these regulations you will need to do a lot of research and make a convincing case.

Using an excavator makes it possible to loosen the soil to a greater depth and to introduce biomass. Unwanted plant growth and roots can easily be removed and worked into the soil. The beneficial effects of this technique are long lasting and the area will not need to be dug over every year. Introducing these plants and roots increases the soil's capacity for water retention and therefore also improves the water balance. In addition to this, the soil is loosened and aerated. Where there is oxygen there is life. Even if you are trying this out for the first time, you will quickly see that this way of dealing with the soil brings the best results. It will give us an area of land we can sow, plant and use to grow crops.

I would like to state emphatically that the practice of removing and burning biomass is a mistake. A large amount of biomass is lost by burning it. The little ash that is left over and used as fertiliser is easily blown away by the wind or washed away by rain. The accumulated material should always be put to use right where it is found. It can be used to construct roundwood shelters, paddocks, and the foundations for raised beds, or it can simply be used as mulch.

Whilst using an excavator, I have often come across layers in the subsoil, which stop the water from draining properly. Loosening these layers up and mixing in sand, stones and humus is a lasting way to make the soil productive again. This helps to ensure good plant growth and a healthy soil.

Some of the larger raised beds should be rebuilt every five to 10 years (depending on the type of cultivation). Beds can be flattened or compressed by animals. Also, if there is too much grass growing on a bed it can become trodden down and it can compact the bed. Using a compact digger, it is very straightforward to repair the beds or to rebuild them. The work can be carried out just as well with a tractor and backhoe loader or plough – this is very similar to the way spargel (white asparagus) beds are maintained. The machinery you use should depend on which method is the most convenient and requires the least amount of energy.

Dealing with the Authorities

The first thing I have to know is what I want and what I need to achieve it. It is only once I have complete faith in my project that I can find a way of making it possible. In my experience, the official organisations – whether they are the local council, the Chamber of Agriculture or any other authorities – rarely give useful advice on farming. From what I can see, practically nothing is allowed at the moment. If I took that seriously my options would be very limited. My ability to think creatively and to innovate would fall by the wayside. I have to know what I want and what I am capable of. These days we need a little imagination and courage to lead a life in harmony with nature.

Thank heavens farmers are still free to do what they want with their land provided they use it for agricultural purposes. Unfortunately, many farmers are misled into thinking that they can only do what the authorities will support them to do. If they want a subsidy, the project will have to be officially approved; project documentation and plans will have to be submitted. Larger projects must be approved for agricultural credit and the bank must perform a profit evaluation. No one that goes down this path of dependence and submission will have much success.

Here is an example of how a government-approved project might go: you want to make an unimposing little pool for bathing with a surface area of around 200m². Maybe you also want some fish or a few geese or ducks; or just to have a nice stretch of water on your land. So you take the normal route and go to the town hall. They tell you that you will have to go to the water regulatory authority to make sure your plans do not conflict with local water law, because the town hall is only responsible for the building side of things. You get the same information from the Chamber of Agriculture. They announce that they will support your project, but only if you can provide plans and project documentation drawn up by a builder or hydraulic engineer. Now the approval process is in full swing. It turns out later on that you need to get approval from the fishing authorities just to keep a few trout in the pond. You will also need to have a stability survey carried out. Last but not least, they have to find out if the pond will affect your neighbours. This means that all of the neighbours will be presented with your plans.

Now one of your neighbours thinks that the pond might pose some kind of a hazard. Someone might fall in and it would only attract more midges. Frogs or even snakes might start to appear. The pond might also encroach on neighbouring land. The number of obstacles in your way seems endless. When faced with all of these problems and red tape many people who want nothing more than to put a pond in a field end up deciding it would be better just to give up. But with just a little spark of creativity you can find another way.

You could, for example, find a part of your land that is naturally wet. You could remove whatever plant growth there is and use it to make low banks. The depression in the ground is not a hazard, because the water is not above ground level. Water will only collect in the depression. The surface of the water will only span a few square metres. In time the banks will knit together. The hollow can also be made a little deeper, which hardly anyone will notice. It will not pose a danger to anyone, because very little has actually been changed.

The resulting wetland can be made even larger using this method. Neighbours and visitors alike will enjoy watching the pond and plant life as it grows and flourishes. Children will tell their parents and get them interested in it. Then maybe they will try to make something similar for themselves. In the unlikely event that a neighbour does go to the authorities to complain about the pond having no planning permission, you can explain to the officials from the water regulatory authority that it has always been there. You have only taken measures to maintain it.

If you ever meet an ecologically minded official, then they will sympathise with your approach and the measures you have taken and simply close the matter. In any other case, the authorities will have to prove you wrong first. If there is any doubt in the matter, it will be assumed that you are telling the truth. You just have to keep on trying. The value of a biotope like this far outweighs all the hard work the bureaucrats demand that we do. Do not be dissuaded from your project by seemingly incomprehensible laws or be daunted by the administrative system. Just think for a moment – and you will be able to find friends to support you in your endeavours.

Setting up a Permaculture System

General Questions

Many people want to manage their land in a natural way or to switch from using traditional farming methods to permaculture ones, and ask me about the best way to make use of their land. I have to ask them some questions of my own first: what is their goal and what do they expect from their land? Do they want to be able to live off of it – whether they just want to be self-sufficient or produce food as a business – or would they rather just have a pleasant place to relax in and grow a few herbs and some fruit and vegetables for the kitchen? Are they interested in using their land for growing plants, keeping animals or agroforestry? Will the area be open to the public as an ornamental garden; somewhere they can pick their own food, or a therapeutic oasis of calm? The answers to these questions are the foundations for success. It is important to do exactly what makes you happy, piques your interest and encourages your thirst for knowledge. Then work will not feel like a chore and success will follow naturally.

People often want to realise the dreams they have had since childhood. It is gratifying to see how happy people are when they finally make these dreams a reality. During the design phase, it is always important to consider the interests of the people involved. If the whole family is enthusiastic about the project then this gives us many more options. Spouses, children and parents should all be able to get involved in the planning and design process. For example, most children will be delighted if they are given a small area of land to experiment with and look after by themselves. You just have to trust yourself and follow your instincts, and then you will be on the right track. You have to know what you want – only then will you achieve independence.

Assessing the Land

Once I have worked out what I want to do with the land, I have to examine the area a little more closely. Soil conditions, elevation, climate, exposure, relief, drainage basins, previous use of land and plant growth are all-important factors to consider when planning a permaculture system.

Aspect and Climate

The aspect (the direction the land faces) and elevation of the land affect the design of the system to a great extent. Obviously, it is far easier to set up a productive permaculture system at low altitudes, on flat ground and in sunny places than on steep slopes or at high altitudes. It requires a lot less energy (i.e. working with diggers and growing plants). But even in so-called 'unfavourable' locations it is possible, with a little skill, to set up a functioning permaculture system.

At high altitudes – from around 1,000m above sea level – I aim to design the system to compensate for the shorter growing season and the lower temperatures. It is important to get the most sunlight and make sure crops will be sheltered from the wind. A windbreak made up of various fruit bushes, fruit trees and flowering shrubs at different heights is very effective. Raised beds also have a similar effect. Wind tunnels should not be allowed to form, otherwise the soil will begin to cool and lose valuable moisture. It is particularly important to take measures against soil erosion on steep slopes. I find that terraces and humus storage ditches as well as ensuring permanent plant cover are particularly effective against soil erosion.

With a little ingenuity it is possible to apply permaculture principles anywhere. Seeds can be sown in cracks, clefts or holes in steep slopes or even rock faces. For example, I planted sweet chestnut seeds in clefts in the rock. Afterwards I filled them up with leaves and sowed broom seeds over the top. To my surprise the most magnificent sweet chestnut trees grew and the broom produced the most beautiful flowers. What happened? The layer of leaves covering the sweet chestnut seeds gave them enough moisture to germinate. The roots found their way from the cleft down into the soil and could even force themselves through the rock. The microclimate helps both the broom and the sweet chestnut to thrive.

Even at high altitudes, south-facing slopes offer a multitude of possibilities for growing vegetables, fruit and berries. There the crops will have enough hours of sunshine to ripen. However, the difference between the temperature during the day and at night is so great that it makes south-facing slopes vulnerable to frost damage. In spring, freezing at night and thawing out again in the day is particularly dangerous for the crops. This makes it especially important to choose hardy varieties. During dry summers, soil on south-facing slopes is liable to dry out. Once again, keeping the soil covered using green manure crops will help to protect cultivated plants. Bare earth dries out quickly and then has no protection against the wind and rain. This results in erosion and the loss of nutrients.

On north-facing slopes and areas with just a few hours of sunshine it is important to choose early-maturing varieties, which can still ripen fully under these conditions. To make the best use of the warmth and sunlight I use many different techniques to capture heat. For example, it is possible to make a niche in the hillside. You should place as many large stones on the hillside as possible. They store the heat like a masonry stove and release it again slowly into the surrounding area. I place plants that need a lot of heat next to the stones. If possible, I put a pond or lake in front of the niche. The sun's rays are then reflected by the surface of the water and the overall effect of the niche is increased. This helps the niche to gather heat and therefore serve its purpose as a suntrap. This way even plants that require very warm conditions can be cultivated at high altitudes and on north-facing slopes.

A sun trap in a niche: castor-oil plants, tobacco, cucumbers, pumpkins, courgettes, sunflowers and many other plants thrive here in a polyculture at 1,300m above sea level.

Soil Conditions

It is particularly important to get a feel for the quality of the soil you will be dealing with. The more I know about the properties of the soil, the better I can work with it. It is vital to make an accurate assessment of the soil if you are going to be reshaping the land. You must identify and determine any risk of landslides. It is also a good idea to find out what sources of water there are. Are there areas of marshy land or places where water has accumulated? What is the soil type? Is it a light, medium or heavy soil? How deep is the soil and how well developed is the humus? I have to answer all of these questions if my design is going to be successful. My ability to assess the land helps me to select the plants that will improve the soil the most. The more fertile the soil is, the more successful the permaculture system will be.

It is a soil's structure that makes it good or bad. The best soil has a crumbly structure. A crumbly topsoil allows plants to establish their roots more easily. Its high pore volume means that it holds water and nutrients like a sponge. The many invertebrates and microorganisms that live within the soil help to create this crumbly structure. One of these creatures is the earthworm. The positive effect earthworms have on soil is well known and the crumbly structure of worm casts is clear for anyone to see.

It is also important to consider the pH value of the soil. This is determined by the soil's mineral composition, but, like most properties of soil, it can be altered by plants and the creatures and microorganisms that live within it. There are plants that prefer an acid soil and others that grow better in an alkaline soil. Most cultivated plants grow best in a slightly acid soil (between pH 6 and 7). A near-neutral pH value is particularly good for the health of the soil, because most microorganisms that live in it function best under these conditions. The more effectively they can work, the faster biomass and humus can be produced. An increase in soil acidity, which is frequently caused by monocultures and the use of fertilisers, leads nutrients to be washed away and the crumbly structure of the soil is lost. This in turn has a negative effect on the balance of air and water within the soil.

• Assessing the Soil

If you want a detailed soil analysis, you can have a soil sample examined to measure nutrient content, composition and pH value. An Institute of Environmental Engineering (e.g. in Graz or Innsbruck) or one of the many private companies will offer these services. In any case, I think it is very important to develop a feel for the soil yourself. There is a tried and tested way to determine the soil type. It is called the 'finger test' and it is very easy to carry out.

To perform the test take some fresh soil (not dried out) and roll it between your palms or finger and thumb. The stickiness and how easy the earth is to mould varies from soil type to soil type. You can also find out how large the grains of soil are in the same way. The first thing I determine is whether the soil is 'light' and made of sand or loamy sand, 'medium' and made of sandy loam or 'heavy' and made of loam, clay loam or clay. The 'weight' of the soil depends on how well the materials it is composed of bind together.

To begin with, I try to roll the earth between my palms to about the thickness of a pencil. If this does not work it means that the soil is sandy. Otherwise I am dealing with at least a 'medium' soil of sandy loam. If I can roll the earth to half the thickness of the previous one, then it is heavy loam or clay. To tell the difference between the two I can break the roll in two. Shiny layers indicate clay, whereas matt layers indicate loam.

• Characteristics of 'Light' and 'Heavy' Soil

'Light' soil is well aerated and heats up quickly. However, its fine grain structure makes its capacity to store water and nutrients low. This means that plant cover is needed at all times. The plants will help to produce humus and prevent the topsoil from drying out. 'Heavy' soil, on the other hand, retains water easily. The nutrient content is higher, because the soil stores the nutrients more effectively. 'Heavy' soil is also poorly aerated, which means it is prone to compaction. Its average soil temperature is lower. It is as hard for plants to establish their roots in it as it is for people to work with it. Raised beds have many advantages when dealing with this kind of soil. Constructing the beds loosens the soil and the introduction of biomass helps to aerate it. Well-aerated soil warms up more quickly and stores the warmth well, because air does not readily conduct heat.

By introducing large stones to store heat, the sun's energy can be harnessed and the average soil temperature will increase. I use small structures, windbreaks, hedges and rows of trees to slow down the wind, which always travels at high speeds. These stop it from carrying all of the heat away, and create a useful microclimate with a higher soil temperature where I can grow crops. The average soil temperature is an important factor for the germination and growth of plants. Even the microorganisms that live in the soil are more active at higher temperatures. Decomposition takes place more quickly and I have good quality humus for my plants in very little time.

• Indicator Plants

The plants growing in an area tell us a great deal about the nutrient ratio, pH value and the general condition of the soil. With a little practice, it is possible to assess the soil conditions based on the vegetation growing in the area. If there are nettles, hogweed or orache then the soil is rich in nitrogen. In this soil I can grow plants that need a great deal of nutrients like root vegetables and tubers. If there is a large quantity of sorrel, the land will be suitable for growing Jerusalem artichokes (*Helianthus tuberosus*) and sunflowers (*Helianthus annuus*), because they take up the excess nitrogen and provide valuable green material, tubers and seeds. In this way I deprive the orache and nettles of nutrients. They are quickly overshadowed by the other plants, which have grown tall, and begin to die.

It is important not to exclude any plants when assessing the soil conditions. You will need as many indicator plants as possible to make an accurate analysis. Certain combinations of plants or an above average number or certain varieties can help you to determine the soil conditions immediately. To give you an idea of this I have made a short list of indicator plants:

Nitrogen rich soil:
Chickweed *(Stellaria media)*
Stinging nettle *(Urtica dioica)*
Annual nettle *(Urtica urens)*
Cow parsley *(Anthriscus sylvestris)*
Hogweed *(Heracleum sphondylium)*
Elderberry *(Sambucus nigra)*
Common orache *(Atriplex patula)*
Goosegrass *(Galium aparine)*
Shepherd's purse *(Capsella bursa-pastoris)*
Fat hen *(Chenopodium album)*
Mugwort *(Artemisia vulgaris)*
Nitrogen poor soil:
Sweet vernal grass *(Anthoxanthum odoratum)*
Sheep's fescue *(Festuca ovina)*
Mouse-ear hawkweed *(Hieracium pilosella)*
Corn chamomile *(Anthemis arvensis)*
Broad-leaved thyme *(Thymus pulegioides)*

Alkaline soil:
Meadow clary *(Salvia pratensis)*
Pheasant's eye *(Adonis aestivalis)*
Forking larkspur *(Consolida regalis)*
Salad burnet *(Sanguisorba minor)*
Betony *(Stachys officinalis)*
Sanicle *(Sanicula europaea)*
Blue moor grass *(Sesleria varia)*

Acid soil:
Sheep's sorrel *(Rumex acetosella)*
Bracken *(Pteridium aquilinum)*
Heather *(Calluna vulgaris)*
Bilberry *(Vaccinium myrtillus)*
Corn chamomile *(Anthemis arvensis)*
Creeping soft grass *(Holcus mollis)*
Wavy hair grass *(Avenella flexuosa)*
Mat grass *(Nardus stricta)*

Dry soil:
Bugloss *(Lycopsis arvensis)*
Whitlow grass *(Erophila verna)*
Broad-leaved thyme *(Thymus pulegioides)*
Golden marguerite *(Anthemis tinctoria)*

Wet soil:
Wood club rush *(Scirpus sylvaticus)*
Purple moor grass *(Molinia caerulea)*
Corn mint *(Mentha arvensis)*
Creeping buttercup *(Ranunculus repens)*
Coltsfoot *(Tussilago farfara)*
Soft rush *(Juncus effusus)*
Compact rush *(Juncus conglomeratus)*

Compacted soil:
Field horsetail *(Equisetum arvense)*
Dandelion *(Taraxacum officinale)*
Greater plantain *(Plantago major)*
Silverweed *(Potentilla anserina)*

INDICATOR PLANTS

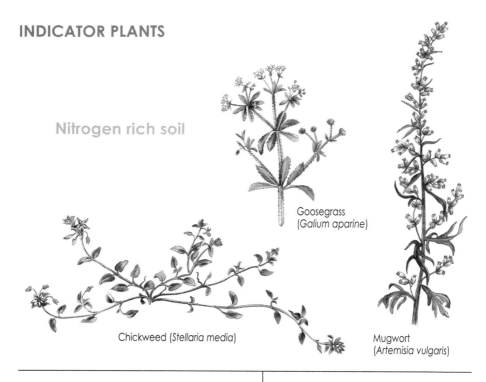

Nitrogen rich soil

Goosegrass
(*Galium aparine*)

Chickweed (*Stellaria media*)

Mugwort
(*Artemisia vulgaris*)

Nitrogen poor soil

Wet soil

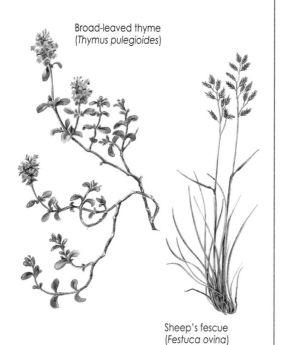

Broad-leaved thyme
(*Thymus pulegioides*)

Sheep's fescue
(*Festuca ovina*)

Compact rush
(*Juncus conglomeratus*)

Creeping buttercup
(*Ranunculus repens*)

Compacted soil

Silverweed
(*Potentilla anserina*)

Acid soil

Mat grass
(*Nardus stricta*)

Sheep's sorrel
(*Rumex acetosella*)

Dry soil

Yellow marigold
(*Anthemis tinctoria*)

Alkaline soil

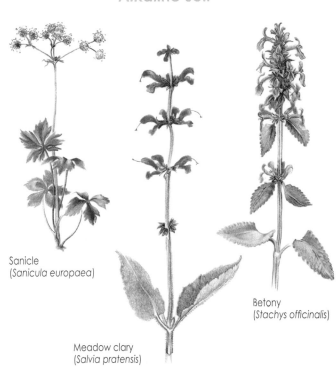

Sanicle
(*Sanicula europaea*)

Meadow clary
(*Salvia pratensis*)

Betony
(*Stachys officinalis*)

• Experiences with Different Types of Soil

In the course of my projects in Austria and abroad I have worked with very different types of soil. When I survey the land for the first time, I dig a number of test trenches in different areas to give myself an idea of the soil conditions. The soil can vary greatly within as little as 10-20 metres, but can also remain the same over large areas. For example, here on the Krameterhof the soil conditions change dramatically within a very small area. At my project in Burgenland, however, the changes were far more gradual. If you are planning to make large-scale changes to your land, whether you want to create a terrace, water garden or pond, you will need to use completely different methods depending on the characteristics of the soil.

On the Krameterhof, the deeper layers (two to three metres) are made of a very distinct coarse and stony material. If I wanted to make a terrace or pond here, I would have to separate this material. To do this I dig out the required amount of earth and shake it to form a tall mound. It is best to use an excavator for this. It can scoop up the earth and then scatter it from as high up as possible. The fine material will stay in the middle, whilst the coarse material will fall to the sides. When I am making terraces or ponds I use the coarse material to support and secure the structures, whereas I use the fine material on the terraces as it makes a fertile soil, or I use it to make the bases of ponds watertight.

Left: Test trench at a permaculture project in Thailand (clay loam).
Right: Test trench for the BERTA permaculture project in Ausseerland (Austria). The layers can be seen clearly (humus, gravel and loam).

Loam requires a very different strategy. If you are planning on building a pond you will not need to separate the materials, except for the humus layer, which should always be separated. With this kind of soil, sealing ponds is not a problem. The challenge is to stabilise the walls of deeper ponds. Loam has a high level of water retention and absorbs water quickly, which means that it easily becomes muddy and takes a long time to dry out. This means that heavy clay and loam soils should not be shaken into tall heaps. Even if you compact it with a digger or a roller, you will not be able to stabilise the soil, because of its high water content. If the accumulated weight of the soil is too much, the increase in pressure will slowly squeeze out the soil moisture.

The separated soil will finally give under the pressure and the earth will sink like an unsuccessful cake. Cracks will begin to form across the surface. When the pond is filled or if it rains, there is a danger that water will enter the walls of the pond. This could lead them to slide. This is why separating soil of this type will only work over a longer timescale. The separated material must have time to stabilise before any further work can be done. Once the soil has had the time to dry out and harden, it will be able to take the additional weight.

Design Ideas

Test Areas

The little gardens I tended as a child were my first test areas. Over the years my test areas grew larger. By experimenting I learnt a great deal about nature. My curiosity never ceased to grow. Now my land measures around 45 hectares, which makes for a very large test area indeed. Although I know very well what will grow and thrive on my farm, I always make a point of sowing new plants. The outcome never ceases to surprise me. Plants that, according to the experts, should not be able to grow here can be cultivated on the Krameterhof regardless. If I had not tried, I never would have thought it possible. For instance, I can grow many varieties of kiwi, lemons and grapes in suntraps.

I cultivate ancient cereals on old pasture at 1,500m above sea level. This is also the result of an experiment. I sowed einkorn wheat, emmer wheat and ancient Siberian grain and, to my surprise, even at this height, they had fully ripened by September. Although Lungau is the coldest area in Salzburg – hence its reputation for being the 'Austrian Siberia' – cereals can be grown here despite the high altitude. Many experts claim that Lungau is not suitable for growing cereal crops and that the higher areas are even less so. Despite this, barley, wheat, oats, rye and even flax and sunflowers ripen fully on our farm at 1,500m above sea level. However, this only works with old hardy varieties. These varieties, unlike the standard EU-approved seeds, can cope with the poorest soils and the most extreme temperatures. The nutritional value and content of the cereals I grow on my mountain pasture is far better than that of cereals grown as a monoculture.

Orchard at 1,400m above sea level: a colourful assortment of different varieties and countless supporting plants stabilise the system.

Many different kinds of fruit bush and fruit tree grow very well at this altitude. Naturally, these cultures do not give the same yields that are possible in lower regions. Scientific research has shown that the nutritional value of many fruits increases when they are grown at higher altitudes. This is mainly the result of the harsh, cold nights, which help to improve the flavour. Growing in these so-called 'unfavourable' locations also makes it possible to offer high quality, aromatic fruit at times of the year when there is very little competition. Specialist distilleries that use organic fruit and berries are particularly enthusiastic. Our products are used to make distillates, juices, vinegar and cider. The price is higher than the usual market price, because the product is of a higher quality. This more than makes up for the lower yield compared to that of fruit grown in 'favourable' locations.

My fruit trees grow at 1,000m above sea level in the wildlife area all the way up to 1,500m on the Lanschütz, which is an area named after the local mountain. 'Kassin's Frühe' cherries ripen fully in the wildlife area by the end of June, whilst they can only be harvested on the Lanschütz at the beginning of September. The situation is much the same with currants, pears and apples. On our farm, the 'White Transparent' and 'Stark's Earliest' varieties ripen in the middle of August at 1,000m, but from the middle to the end of September at 1,500m. At 1,100m above sea level the 'White Transparent' is so floury by the end of August that it cannot be used for juice or cider any more. At 1,500m above sea level, however, it is still an excellent juice and cider apple at the end of September.

I have also introduced mushrooms into my experiments. According to experts, shiitake mushrooms can only be cultivated at lower altitudes, because they need a great deal of warmth. As an experiment, I inoculated a 50cm-thick oak log with shiitake mushroom spawn at 1,500m above sea level. Afterwards, I sunk the log 30cm into the ground and the same way up as it would have been growing, to provide it with the necessary water. Two years later the first mushrooms appeared, but there were so many of them that the entire log was covered. Mushrooms continued to appear every now and then until the first frosts. It has been more than ten years and the log is still producing new mushrooms each year. In the last few years shiitake mushrooms have even started to come up through the earth around the log. For ten years I have done nothing to help the mushrooms grow, I have merely harvested them.

After this successful outcome, I decided to try growing mushrooms on living wood. I bored holes in a number of broadleaf trees with a hand brace and inoculated them with mycelia. Naturally, I used only one kind of mushroom per tree. Unfortunately, the experiment was not successful. The trees rejected the mycelia and the holes healed over. However, mushrooms appeared on the ground around the trees that had been ring-barked before being inoculated and had bark on the earth around them as a result. Any crop from this method would, admittedly, be somewhat limited, as the trees would die within a year and be quickly blown down by the wind or collapse under snow.

Test areas are very important and you should never stop experimenting. There are many more things that are possible in nature than you will find written down in books. However, you will only discover this if you are ready to fail and you are ready to learn.

Microclimates

Microclimates are a very important aspect of any permaculture system. Every microclimate forms a special biotope that is colonised by a particular community of plants. A large number of animals find food, a habitat, somewhere to breed and take refuge; the different microclimates give useful insects somewhere to breed as well. This is why I try to create as many microclimates as possible throughout a permaculture system. The diversity of plants and animals helps to create a system in which every species will find its natural balance. This is the only way to prevent the population of any one species from becoming dominant and reducing the overall integrity of the ecosystem.

Microclimates are areas in which the climatic conditions are completely unlike those of the surrounding area. This means that these places can be comparatively dry, wet, shady or sunny – all depending on what the microclimate is needed for. This allows me to create the correct conditions for very different kinds of plants in a relatively small area.

Microclimates may develop naturally near large stones, in clefts in the rock, in hollow tree trunks, near tree stumps, in hedges or amongst trees and

Microclimate on a rock face: spaghetti squash growing on the rock.

shrubs. Favourable climatic conditions can also be created in an area by making terraces, raised beds and ditches. It is especially important to keep the lines of the terraces and paths as rounded and winding as possible. Straight lines create wind tunnels, whereas curved shapes make niches. These niches are sheltered from the wind and can work as suntraps. In particularly exposed places I also make hills and hollows to lessen the effects of the weather.

On my travels in South Africa, northern Brazil and Colombia, I have seen large stretches of land lying fallow. The land is completely unprotected against erosion and is gradually drying out. It is in exactly these countries that we need to use microclimates to change unproductive areas back into fertile land. On this kind of land I could, for instance, plant particularly hardy and fast-growing trees to create a kind of 'pioneer forest' to protect the land from erosion and prevent it from drying out. Later on, more demanding fruit trees can grow safely amongst the initial ones. Once they are stable and have grown large enough, the pioneer trees will no longer be required and can be cut down and used for timber. Another way to establish cultures is by making raised beds on top of large branches and shrubs. Then I can introduce seeds like mango or papaya into the bed. Although different crops like manioc, or the seeds of trees that produce valuable wood can be sown as well. The seeds will probably lie dormant for some time. At first they will not encounter the conditions they need to germinate, because it will still be too dry. However, once the rains come water will begin to collect. The biomass within the raised bed will retain water and slowly begin to decompose, allowing the seeds to germinate.

A mulch of leaves and straw can be used to stop the germinating seeds from drying out. Laying acacia branches or any other thorny branch in the newly created microclimate will protect the choice plants from being eaten. I have to make these places as inhospitable for the animals that eat the plants as I can. If possible I lay an entire thorn bush or tree on the bed. It will wither, it is bulky and it will keep the animals away. It protects the plants and also slowly rots back down into fertiliser.

At the same time I also sow many plants that the animals prefer, because this is the only way to protect the plants that I do not want to be eaten. The bulky material used to protect the plants has a further advantage: fine material carried by the wind is deposited on the beds and a small biotope begins to develop. This is how useful systems, which retain moisture, protect the soil from erosion and prevent the plants from being eaten are created.

In the Scottish Highlands the situation was completely different: the areas that I visited had been cleared centuries ago. Now there is nothing but miles and miles of heath with not a tree to be seen. The rainfall there is very heavy and the wind is quite strong and never stops. This makes it difficult for anything other than heather and sedge to grow. The pH value of the soil has sunk to a value between four and five, so any yield from such an area would be minimal. In places like these it is vital to set up suntraps and windbreaks.

As there are so many stones, small stone walls and islands can be constructed. In the lee of the islands there is shelter from the wind and the stones balance out the temperature. The wind also deposits fine material and a humus layer gradually begins to build up.

Between the stones, I planted and sowed different varieties of willow, wild rose, broom, lupin, sweet clover and comfrey as pioneer plants. The permaculture system in Scotland is now getting on wonderfully. Now that the first small-scale attempts have been successful, it is possible to create a larger-scale biotope, possibly using a mechanical digger. As the system develops, the land will begin to resemble a rag rug. Many irregularly shaped landforms will be created, which will produce numerous microclimates. In this way the diversity of plants will continue to increase.

When you are creating terraces and raised beds to improve the microclimate, you should take the existing climatic conditions into consideration. In places that are rainy and windy you will need to do the opposite to what you would try in hot and dry places. For example, in the Scottish Highlands I made sure that there was always drainage in place to take the

Protected by a larch trunk, even bitter oranges (*Poncirus trifoliata*) can flourish.

excess water. If I had not done this, the raised beds would have become acid. In drier areas the water must under no circumstances be drained away, instead the land should be designed to retain it.

As plant material is broken down inside the raised bed, heat is released and this helps to encourage healthy soil life. Choosing the right plants will lower the acidity of the soil and allow a greater number of plants to grow on the heath. During my experiments in Scotland, it became clear to me that we would need higher fences to deal with the increased danger of plants being eaten. Even the black and willow grouse could fly over the then two-metre-high fences around the test areas and eat the plants and seedlings.

The beneficial effects of microclimates have even allowed me to grow cacti (prickly pear; *Opuntia ficus-indica*) outside during the winter, and apricots (*Prunus armeniaca*), sweet chestnut (*Castanea sativa*), grapes and kiwi fruit (*Actinidia deliciosa*) in particularly warm and sheltered areas on the Krameterhof. As these plants are so sensitive I also make sure that a blanket of leaves from nearby trees protects them during the winter.

Terraces and Paths

Human beings have known the benefit of using terrace systems for a very long time. In Asia, South America, Africa and Europe people have been using terraces to cultivate cereals, vegetables, coffee, tea, herbs and grapes for thousands of

Newly-created terraces in Burgenland. A considerable area of extra land has been gained on this south-facing meadow.

years. Making steps in steep slopes helps to prevent soil erosion. Valuable humus remains on the slope instead of being washed away. Terraces store and hold moisture so that plants have access to rainwater and meltwater for longer. Terraces increase the area available for cultivation, are more pleasant to work and are far easier to access than a steep slope. You can stroll along the terraces and just take in your surroundings. The number of ways in which the land can be used will increase and its value will rise. Well-designed terraces minimise the danger of landslides and mudslides and also greatly improve the microclimates in cultivated areas.

I tend to design terraces so that they can be used as paths and provide access for mechanised farming equipment. By combining the two uses, I have both a terrace that is a path and a path that is a terrace. The two possible uses are always open to me. Of course, this will only work as long as no single terrace is used as a path for too long, otherwise the soil will become compacted and the crops will suffer. It is still possible to access the terraces while they are under cultivation, but it is important to keep to the embankments.

Building a Terrace System

• Width

Before any terraces are created, it is important to think about how they will be managed. The width required by any machinery you are planning to use should also be taken into account, so that working the terrace will be straightforward and the crops will not be damaged. It is best to make sure that the entire terrace can be cultivated or harvested in a single trip. This way the least amount of energy is used and the damage caused by machinery is minimal. In my experience, managing two terraces each with a width of five metres is far easier and more profitable than managing a single terrace with a width of 10 metres. Creating a narrower terrace also requires far less earth to be moved. When you are calculating the perfect width for a terrace, it is important to consider the current gradient of the slope. The steeper the slope, the narrower the terrace should be. The shallower the slope, the wider the terrace. The prevailing soil conditions should also be noted. Particular care should be taken with fine, loamy soil on very steep slopes, because this is where there is the greatest danger of erosion. Under these conditions I would only make very narrow terraces.

• Gradient

The gradient of the terraces depends on the accessibility and development of the other plots of land. The gradient should be as low as possible and no more than 15 to 20 percent. The terraces should be laid out to make as much of the land accessible as possible. It is a good idea to have shortcuts and paths to connect the terraces, so that you will not have to travel the full length of the terrace when

Terraced landscape on the Krameterhof.

you want to maintain the land. Dead ends waste time and energy and should be avoided. The gradient of the terrace embankments can be 1:1 if the soil is stony. On sandy or loam soils I have had the best results with gradients of around 1:1.5 to 1:2.

• Risers and Separating Material

To stabilise the embankments risers need to be constructed. The topsoil and the layers of earth beneath should be removed until you reach solid and stable material. The riser forms the foundation of the embankment and is angled slightly into the hill. Next, the embankment should be built up replacing the excavated layers. The topsoil forms the uppermost layer.

This work can be carried out very easily with a mechanical digger. A specialised excavator can fit into the smallest spaces; a mini digger can get through a garden gate and a walking excavator or 'spider' can even climb over a fence. When the correct equip-

The walking excavator ('spider') can even work on rough terrain.

RISER

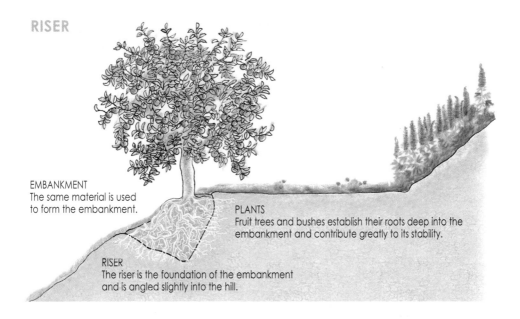

EMBANKMENT
The same material is used
to form the embankment.

PLANTS
Fruit trees and bushes establish their roots deep into the
embankment and contribute greatly to its stability.

RISER
The riser is the foundation of the embankment
and is angled slightly into the hill.

ment for the task is chosen, the work goes quickly and any unnecessary annoyance is avoided.

It is important to handle the topsoil very carefully. Mistakes are frequently made when the operator of the mechanical digger or the person in charge does not have enough experience and fails to preserve the topsoil. Everything is ploughed through, the material is not separated and the risers are forgotten. The material is strewn around all over the place, which naturally makes it very difficult to plant up later on and also increases the danger of landslides.

• Working with Water

In areas where there is little rain, I try to angle the terraces into the hill slightly to help retain water. When stabilising the terraces and securing them against heavy rainfall, it is important to make sure that their alignment will not channel the water, otherwise this will cause significant damage. With a loamy subsoil it is particularly important to manage water well. You should under no circumstances angle any watercourses or ditches into the hill, because the highest terrace can become waterlogged and this greatly increases the danger of landslides. On heavy soils, surface water must be drained off over a large area. This is best achieved by creating the terrace with a slight downward angle. This way you can create watercourses through which large quantities of water can drain away safely.

When putting in paths and roads, it is common to dig ditches and install culverts at intervals of 50-100m. Unfortunately, little notice is normally taken of whether there is any water present or not, how the nearby subsoil will react to an increase in water or how well the plants can cope with waterlogged conditions. This careless attitude causes vegetation to be killed off by waterlogging and can eventually lead to mudslides and rockfalls.

In my opinion, it makes far more sense to disperse the water wherever it might accumulate – by making the middle of the path or road slightly higher than its edges. This will not lead to channelling and the water can travel at its own speed without causing any damage. Of course, this will not help with streams or springs, which should be diverted underneath paths and roads through pipes or culverts. After the brief diversion they can return to their natural course.

Through dry, stony or sandy soil water will percolate over greater distances. Ditches are particularly good for raising the moisture level of the surrounding soil. They store surface water and runoff and let the moisture seep into the soil. This provides very good conditions for the neighbouring plants. The ditches also collect organic material, providing water and a habitat for many living creatures, and therefore boosting the population of useful animals and insects.

• Stabilising the System

Large stretches of continuous land should not be altered during the growing season, because this increases the danger of landslides. This is why I carry out large-scale projects gradually over a long period of time. In the first year, terraces are made at the top, middle and bottom of the slope and then planted. In the

New terrace system on the Krameterhof: the terraces are around four metres wide. A variety of hardwoods and fruit trees are planted on the embankments. A mixture of seeds (mustard, flax, comfrey and potato) are already beginning to grow.

Within a year of the terrace being built, valuable biomass is already being produced and most of it remains on the surface. A monoculture of spruces used to grow on what was once poor and acid soil. The terrace has quickly made it possible to cultivate more demanding plants.

second year, more terraces are created between the original terraces once they are completely stable. On steep slopes the first terrace should be started at the lowest point. Then you should work your way up. If material begins to slide down while you are working, the terraces below will collect it. The material can then be incorporated into the soil. Stones can be placed on the terrace for additional stability and heat storage.

Choosing the correct plants makes a large contribution towards stabilising a slope. Plants with root systems that grow to different depths are very useful. Once work with the excavator is finished, new terraces should be sowed and planted immediately, because at this point the danger of erosion is at its greatest. The soil is also very loose and moist just after the terrace is created and this provides seeds with the best conditions for germination. When it rains, the seeds will be pushed through the loosened humus layer and further into the earth. Then the

A pond with steps (very narrow terraces) in Burgenland. The water level can be altered to any height. The terraces can also be flooded when required. Evaporation during the summer creates a beneficial microclimate. In the hot, dry summer months (Pannonian climate) of southern Burgenland this is a real advantage.

soil can be covered in leaves or straw. Mulching helps to retain moisture while the plants are taking root and it provides the crops with additional nutrients.

To ensure that the trees and plants root well, it is important that the embankments are very stable and built up using loose soil that is rich in humus. My method differs from that of conventional terraces, where the steep embankments are constructed of heavily compacted soil and then smoothed flat. Seeds are then easily blown away by the wind or washed away by the rain. They also have more difficulty germinating and taking root in these flattened areas of soil.

Managing a Terrace Culture

Terraces can be used to grow any conceivable crop; they can be worked just like fields. It is important to cultivate and maintain plant cover as soon as the terraces have been constructed. If the topsoil is good enough, demanding plants such as vegetables or cereals can be grown straight away. Otherwise, green manure crops will be needed and the soil will have to be prepared before crops can be cultivated. Meadow flowers are also very good for plant cover. If there is a wildflower meadow nearby that has not been cut back for a long time, you will find more than enough seeds there.

You can also add sweet-smelling plants and medicinal and culinary herbs to the mixture of seeds, to create a lush flora. On poor soils or on steep slopes deep-rooted green manure crops like sweet clover and lupins are best. They stabilise

A variety of fruit trees and rowan trees with lupins to improve the soil on a terrace embankment on the Krameterhof.

the terrace with their deep root systems. They also improve the nutrient content of poor soils with their ability to fix nitrogen and make it available to other plants, which is assisted by symbiotic bacteria. In wetter areas Alsike clover can be sown and white clover and black medick can be used for plant cover. Other plants that are suitable for plant cover can be found in the 'Green Manure' section.

Even in the first year, plant cover helps to create substantial amounts of biomass for the terrace culture. These plants produce humus, which continues to improve the fertility of the soil. Using this method, I have managed over the course of time to cultivate the most demanding types of vegetable on what was once the acid soil of a former spruce culture. By autumn the plant cover will have begun to decompose and will protect the soil from frost. The earth will not freeze as quickly, which means that the invertebrates and microorganisms in the upper layers of the soil will survive longer in the spring and autumn. The practice of cutting grasses back in the summer and autumn and removing them to make hay is a terrible mistake. We must discard the concept of order that so many people embrace today and recognise that 'untidiness' is a part of nature.

Once the soil is fertile enough, the crops can be planted. The embankments between the terraces provide relatively dry and warm conditions, which I always bear in mind when planting crops. I have had the best results planting fruit bushes and trees on these embankments. If the right varieties are selected, the fruits and berries ripen in the autumn after the vegetables and cereals have been harvested. This use of seasonal crops makes efficient use of the land and avoids the risk of the crops being damaged. When I am selecting trees and shrubs, I choose varieties that will be useful to me and that can deal well with the local conditions.

Humus Storage Ditches

When making any changes to the terrain, especially when creating new terraces, I dig ditches in appropriate places to hold humus and water. These ditches collect any surplus water from heavy rainfall or snowmelt. They are dotted throughout the entire permaculture landscape. I make them long and wide with low banks, so that the water can be absorbed over a large area. The terraces and raised beds below will steadily be supplied with water. Great care should be taken with ditches on heavy soils: the danger of landslides is at its greatest! It is best to start on a small scale and observe the system closely.

The sides of the ditch should slope gently upward and it should be set well into the hill. As I have previously said about making terraces, it is important not to let the water form channels, otherwise it will cause a great deal of damage. The gentle sloping of the sides helps to prevent this. If the hill already has any hollows or depressions in it, an excavator can be used to make them into ditches very easily. To do this, the excavator uses a two-metre wide 'slope bucket' that

HUMUS STORAGE DITCH

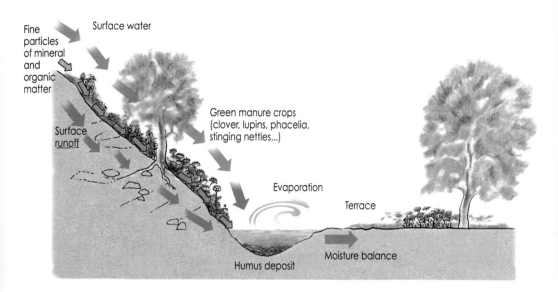

Fine particles of mineral and organic matter

Surface water

Surface runoff

Green manure crops (clover, lupins, phacelia, stinging nettles...)

Evaporation

Terrace

Humus deposit

Moisture balance

can be hydraulically operated in any direction. In this situation it is not normally necessary to dig a ditch. You can just use the bottom of the bucket to push the soil down, making the hollow deeper.

Some ditches are filled with water the whole year round, whilst others, depending on the location and size of the ditch, dry out periodically. The advantage of these ditches is that valuable nutrients and humus will be collected with the surface water when it rains heavily. Once the water level has sunk again, it is easy to extract and use this material on new systems, embankments and crops. This way the best humus, enriched with nutrients, is provided for lush plant growth.

Using ditches in this way also has a very positive effect on the hydrology of my land. The accumulated water evaporates slowly and brings significant long-term benefits to the nearby vegetation. This reservoir of water is vital for the survival of my plants in dry areas and during hot summers, because they do not receive any additional watering. The many beneficial effects of humus storage ditches mean that they play a substantial part in preserving the natural balance of the entire permaculture system. They take very little effort to create and are very useful when managing the land.

Raised Beds

Raised beds have a substantial advantage over normal beds that are at ground level. They create microclimates, which according to their position, relative to the course of the sun and the prevailing wind direction, provide very different plants with the conditions they need. The beds are built loosely, which helps the soil to retain more water, and they soak up rainwater like a sponge. The water is stored in the lower levels of the beds and the hollows between them, while the raised part dries out far more quickly. The result is both dry and wet areas. It is also my experience that the raised part of the beds warms up more quickly, which is a great advantage in colder climates and at high altitudes. Well-aerated and correctly-planted raised beds can help to slow down the freezing of the topsoil. If the beds are made of organic material, the inner part of the bed will slowly begin to decompose. This releases heat, which in turn improves the conditions for germination and plant growth. The decomposition also releases nutrients, which makes it possible to cultivate more demanding varieties of vegetable without using fertilisers. The shape of the raised bed provides a larger area for cultivation. On small plots of land – like town gardens – gaining this extra space is particularly important. Finally, building raised beds offers many exciting possibilities for garden and landscape design.

Raised bed with terraces forming a crater garden.

Design Ideas

In a number of gardening books it is becoming increasingly common to find instructions for making raised beds. Most of them give dimensions for the perfect raised bed to the centimetre. These kinds of detailed instructions make it easy to construct the bed exactly as it is described. Free thinking and creativity are quickly lost. There is no template for the perfect raised bed in Holzer permaculture, because the beds can have very different dimensions. During the planning stage, I consider the local conditions and the individual requirements of the people that will be managing it. Although the beds can vary considerably in their dimensions, all of them create the positive effects that I have already described.

The beds vary in height, length, width and shape according to function, location, soil conditions and the preferences of those involved. Flat areas of land in particular offer a variety of interesting aspects to experiment with: the beds could be made in wavy lines of different heights, they could form a half moon, a maze or a circle. In the centre of the circle there could also be a pond. In Burgenland, for example, I made a crater garden. In the summer, a beneficial humid microclimate develops in the crater. This provides a number of very interesting possibilities for cultivating plants.

Even the way the foundations of my raised beds are made varies to reflect local conditions. I do not think it is necessary to state exactly how the layers should be arranged or what material should be used to make the foundations. It makes the most sense and is far more economical to work with the material that is already to hand.

For many years, I chipped branches, shrubs and trees and mixed them with earth to make raised beds. This made for very exhausting and laborious work. Eventually, I tried making a raised bed without chipped material; instead I incorporated thick branches and entire shrubs into the bed. This bed gave a far greater yield than I had expected. The reasons for this were obvious: when spreading the chipped material I had to be very careful; I could not incorporate too much wood (no more than a fourth of the material) into the bed. I also had to make sure that the material was spread very loosely, so that it would not compact.

Substances like resin can also be released too quickly into the earth and the pH value of the soil sinks. In the worst cases the soil acidifies and the yield suffers. I found that introducing much bulkier material had exactly the opposite effect. Although the raised beds tend to be much larger and higher when entire trees are incorporated into it, the aeration of the system is vastly improved. The bulky material causes small shifts to occur throughout the bed as it slowly breaks down and as it responds to changes in the moisture content of the soil. It contracts and expands again, which keeps the structure of the bed loose.

Bulky material rots down more slowly, which lessens the danger of the soil becoming acid or of the crops being overfertilised. Tree trunks are also excellent at maintaining a balanced level of moisture within the system. This kind of raised bed is particularly good for growing potatoes and other root vegetables, I have also used them to cultivate cereals. I have had good results using these beds in spruce forests as well. Raised beds of this type can last for ten years or more without any major rebuilding, which is much longer than ones made with wood chips.

The consequences of misguided forest management: the storm brought down more than three million solid cubic metres of spruce trees in Austria!

In November 2002, my simple method of building raised beds came in useful again when large areas of the extensive spruce monoculture in Lungau suffered storm damage. The heavy winds caused serious damage to the monotonous forests surrounding the farm. Today, they are still dealing with the fallen trees. On the Krameterhof the damage was minimal. The only victims of the storm were some small stands of spruce that were awaiting official permission for clearing and recultivation. A few of them fell onto my fruit trees and fences.

My plants happily withstood wind speeds of up to 170km/h. I incorporated the fallen spruces I found into my raised beds just as they were. As the opportunity was there, I decided to build a couple of open shelters and new paddocks for my pigs out of the remaining wood from the fallen trees. It is always better to make the best of a situation instead of just complaining.

Wood from trees blown down by storms usually fetches a low market price. When there is suddenly so much of it, it is normally very difficult to get a good

On the Krameterhof, we incorporated the fallen spruce trees into a number of structures. They were used on land that we bought from Austrian Federal Forests in 1988 which has not yet been recultivated. Shown in the photo: spruce tree trunks used to terrace and build a raised bed.

Another way to use the tree trunks: building new paddocks and open structures for pigs and cattle. The structures can also be used for shelter, storage or to grow mushrooms.

price for the wood at all. In addition to that, a great deal of useful timber is lost, because the tree trunks have snapped in the middle. The large number of fallen trees makes it dangerous as well as expensive to get vehicles in to remove them. Frequently, the cost of clearing the trees is greater than the money made by selling the wood.

These examples should make it clear that creativity and imagination are what you need most to build raised beds. The way you organise your land is entirely up to you. You only need to make sure that it fulfils its purpose and that the areas which require harvesting are easily accessible. It is also a good idea – if possible – to build at least two raised beds next to each other. In the hollow between the beds moisture is retained for much longer, which is very useful during hot summers.

Designing a Raised Bed System

Before you start building a raised bed system, you should find out what direction the wind usually comes from and take note of it. The simplest way to do this is to tie a strip of material to a tree or pole and observe it regularly over a period of time. You should also check it at night. This way you can find out very quickly

RAISED BEDS ON SLOPES

CORRECT

The raised beds are at an angle to the slope. The beds will be evenly distributed with water. The water can be absorbed and retained easily – there is no danger of channelling. Image p.65 (bottom)

INCORRECT

Parallel with the slope: the raised beds at the top will receive an excessive amount of water (danger of landslides), whilst the lower ones begin to dry out.

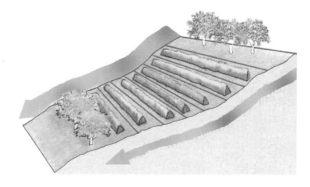

DESIGN IDEAS FOR RAISED BEDS

Course of the sun
(East-West)

Wind direction

A raised bed used as a windbreak, to keep out pollution and as a visual barrier, which can also be harvested; plants in a polyculture with flowering shrubs and fruit bushes. The raised bed is stabilised by deep-rooted plants.

The raised beds in the middle of the system are positioned to catch as much sunlight as possible. The plants are in a polyculture (here with vegetables). Herb spirals make an excellent addition to the other plants and make the best use of the available space.

which direction the wind comes from and which areas are the windiest. If necessary, a windbreak can be put up around the system or the entire system of raised beds could be positioned against the wind and used as a windbreak itself. I find that raised beds planted with fruit bushes and tall-growing plants like sunflowers, Jerusalem artichokes, or hemp make the best windbreaks. I build these beds to a height of at least 1.5 metres. They are exactly like normal raised beds, except that I make the sides a little steeper. This way the beds will not compact so quickly under the increased pressure. With raised beds that are higher than three metres, I put a narrow terrace on the top. This makes managing and harvesting the bed easier. The higher the bed is, the more space will be taken up and you will need to allow for this in your plans. Raised beds not only make good windbreaks, but also make excellent visual barriers and keep out noise and pollution. Frequently, it is enough just to have these windbreaks surrounding the system. I can also angle the beds to give them more sunlight. On steep slopes this is not so easy, because you also have to take into account where the surface water drains.

With raised beds on hills it is very important to pay attention to the flow of water within the system. The beds must not be parallel to the slope, otherwise those at the top of the hill will absorb all of the water when it rains, whilst the beds at the bottom will, in the worst case, begin to dry out. Water must be supplied evenly to all of the beds. The water must not be allowed to channel either or it could lead to landslides. The alignment of the beds in relation to the hill should be determined by the course rainwater takes down the slope.

A system of raised beds can be built by hand or with a mechanical digger, although only relatively small material can be incorporated into the beds when they are built without using machinery. As my experience has led me to favour bulky materials for constructing raised beds, diggers are indispensable for me. I use the digger to make a ditch 1 – 1.5m deep and around 1.5 – 2m wide. I carefully remove the humus layer and separate it. Then I place shrubs and trees along with their roots into the ditch. On top of that I loosely heap a mixture of earth, fine organic material and turf. Finally, I take the humus that was removed and place it over the bed.

If there are no trees or shrubs to use for the bed, I have to make do with turf. Having additional organic material brought in from elsewhere would waste far too much time and energy.

The sides of the raised beds should, depending on the material, be at an angle of at least 45 degrees. I have had good results with even steeper beds of 60 to 70 degrees on heavy loam. Even with a bed made entirely of earth, a steeper angle makes sense. With some materials it is necessary to heap the earth as steeply as possible, as high as it can be and still hold together. When I am visiting other farms or giving advice, I see far too many raised beds that are much too flat. They ask me why the bed is not growing as well as they had hoped. The answer is simple: the angle of the sides is too shallow, so the beds become compacted. The supply of oxygen is decreased, the process of decomposition is interrupted and, if not dealt with, a foul-smelling anaerobic sludge can build up, which has a negative effect on the plants. In addition, the plants will not be able to establish their roots properly, because the ground is too compacted and they will begin to wilt. People continue to make raised beds that are too flat, which makes it all the more important for me to emphasise this point right now.

With wet, heavy soils it is a good idea to put in a drainage system to prevent water from building up. A French drain can be used to do this. With dry and sandy soils, on the other hand, it is important to keep water within the raised bed for as long as possible. This will happen automatically without any additional water being diverted, as it will collect naturally in the hollow between two beds and in the centre of the bed as the bulky material rots down. Covering the surface of the bed with mulch will also stop plants from drying out when they are taking root and are vulnerable.

When the seeds have been sown and the plants are developing, keeping the soil covered will stop them from drying out too much. Crops that are not harvested and other self-set or wild plants can be left on the bed as mulch, which will develop slowly into a rich layer of humus. Having deep, coarse humus and keeping the soil covered are the best ways to retain moisture.

The height of the beds depends on personal preference. I usually create beds with a height of between 1 – 1.5m. This allows people of average height to harvest the beds without difficulty.

BUILDING RAISED BEDS

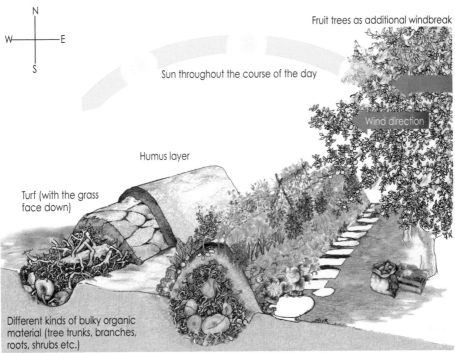

N
W —|— E
S

Fruit trees as additional windbreak

Sun throughout the course of the day

Wind direction

Humus layer

Turf (with the grass face down)

Different kinds of bulky organic material (tree trunks, branches, roots, shrubs etc.)

A stone path. Plant cover, possibly made up of various kinds of clover and thyme that is suitable for walking on, makes for a pleasant path to harvest the bed from.

Managing Raised Beds

It is best to sow and plant raised beds as soon as they are created. As the soil has only just been piled up it is still very loose and has not yet begun to settle. Plants find it easier to establish themselves and take root in loose soil. Seeds fall through the loose soil and are not blown away by the wind so easily. The rain will not wash them away, but instead wash them further into the bed. So as not to hinder this effect, the beds should not be smoothed over. If you are planning to successively crop vegetables and fruit bushes, you should, if possible, plant the bushes on top of the raised bed. The vegetables below can then be reached quite easily. Organising the crops in this way is a particularly good idea in warm, sunny climates, on dry soils and when cultivating plants that need partial shade. Selecting which fruit bushes to use and the intervals at which they are planted allows you to regulate the amount of shade. It is also possible to combine them with fruit trees if you want the whole system to be in shade. Fruit trees and bushes can also be planted between the beds.

The distance between the individual beds can be altered to suit what is being grown. When you are designing a raised bed you should always take into

Raised beds on the Krameterhof in winter.

account how you are planning to manage the bed and what equipment you will be using to do this. Otherwise there may be some unpleasant surprises later on. For example, if I want to use a tractor to harvest the fruit, I have to allow enough space for a path between the beds for the tractor to travel along. This path could, for example, be planted with different varieties of clover for plant cover.

Raised beds are suited to growing all kinds of vegetables: peas, beans, salad, tomatoes, radishes, cucumbers, carrots, courgettes, pumpkins, potatoes and many others. The material breaking down in the centre of the bed provides the plants with plenty of nutrients and the plant growth will be lush. The amount of time the nutrients last or how quickly they are used up depends on what the centre of the bed is made of. If a raised bed is made of chipped wood, which breaks down quickly, a large amount of nutrients will be released in the first year. To make the most of this I select plants that demand a very high nutrient content: pumpkins, courgettes, cucumbers, cabbages, tomatoes, sweetcorn, celery and potatoes to name a few. In beds like these it is better to cultivate less demanding plants like beans, peas and strawberries after three years. If they are planted any earlier they might become overfertilised. Overfertilised plants do not develop a good flavour. With some plants – e.g. spinach – nitrates can also build up in the leaves of the plant, which can be dangerous to ones health if eaten.

Raised beds constructed with bulky material such as whole tree trunks do not develop a particularly high nutrient content in the first year. The bulky material rots down very slowly. However, the supply of nutrients will be steady for many years and there is hardly any danger of overfertilising within the first year. To use a raised bed in the most effective way, you should take into account the nutritional needs of the plants.

I deal with any unwanted plants as I wander around the farm. I simply pull them up and leave them there with their roots facing up. If the weather is very dry and it is around midday, then this is even more effective, because the plants dry out and do not take root again. Mulching, in other words spreading straw, hay, leaves or similar organic matter, is a good way to keep these unwanted plants in check; it also keeps the soil covered and retains moisture.

From the second year, pigs can be allowed on the raised beds for a little time to graze after the harvest. As they search for food, they will till the beds and leave manure. The best fruit and vegetables should be harvested, but enough should be left for the pigs. They should have something to motivate them and make them happy. If too many pigs are allowed to graze in a small area, they can do a great deal of damage. The number of pigs and the amount of time they are allowed to graze must be determined by the available space. When they have worked the soil, it is in the perfect condition for sowing.

Depending on the weather and how they are used, the raised beds flatten gradually over the years. They are then either rebuilt or replaced.

Pick-Your-Own

Many people are beginning to think about the quality of their food and where it comes from. The trend for buying the cheapest possible food is waning and now people want to buy food that is organically grown. The market has picked up on this quickly and has produced many items with the word 'organic' on them and developed new 'organic brands'. The fact that not everything, which has the word 'organic' on it, is actually organically grown is now well known. This is why many people now want to be able to harvest organically grown food for themselves, especially if they can combine it with a pleasant day out.

An appropriately designed raised bed system can make an excellent and relaxing pick-your-own area. By harvesting food for themselves, the visitors feel connected with nature and can convince themselves of the high quality of the produce. This also has many advantages for the farmer: no additional work is required to harvest, clean, transport and store the produce. As everything that is harvested also has to be paid for, any loss caused by having to store unpurchased produce is avoided. The visitors will usually take more than they originally intended to as they wander through the pick-your-own area and see how wonderfully everything grows. Many people start making juices and jams even though they have no gardens of their own. As the visitors see that they are getting genuine organically grown produce, it will fetch a good price.

Raised beds are especially suitable for pick-your-own areas, because the shape of the beds makes it very easy to lead the visitors along a designated path. I cultivate the plants and fruit that I want to offer on long raised beds running parallel to each other. The beds could also form a circle or a spiral. It is

PICK-YOUR-OWN

One possible design
is a system where
high quality food
can be harvested,
combined with a
pleasant day out for
the whole family.

Half – way through there is a large rest area where
visitors can stop for a while to enjoy the water garden,
and nearby nature and adventure playground.

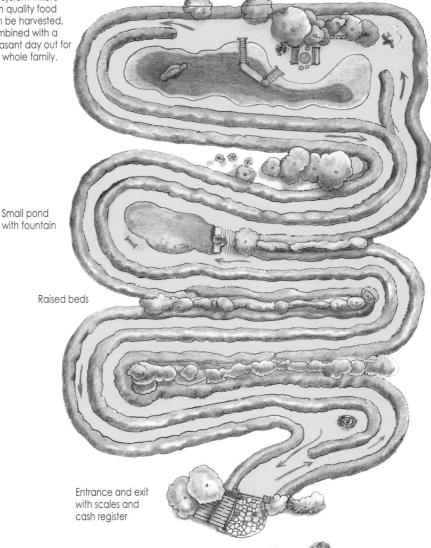

Small pond
with fountain

Raised beds

Entrance and exit
with scales and
cash register

Steep raised beds are at the
perfect height for visitors to harvest:
for children and adults as well as
people who use wheelchairs.

a good idea to make the sides of the beds relatively steep, so that visitors cannot climb over them. Children will also find it easy to reach and harvest the lower half of the beds. Even people who use wheelchairs can easily harvest fruit and vegetables from where they are sitting. Adults can harvest the upper half of the beds without having to bend down. I find it best to place the beds close enough together so that people have to move in single file. At the end there could be an area where the produce is weighed and paid for. Customers need not be charged for what they eat on their way through the pick-your-own area. When running a pick-your-own it is very important to have signs up clarifying that everything harvested – apart from the food eaten along the way as just mentioned – must be paid for. Children in particular tend to harvest far more than their parents can possibly use.

A trip to the pick-your-own becomes an appetising walk that can be experienced through all the senses. Designing the area leaves plenty of room for new ideas. For instance: about half way through the pick-your-own there could be a rest area next to a water garden; a good opportunity to take a break. Near the rest area there might be a herb spiral where the visitors can pick fresh herbs to have with their packed lunches. The children can tire themselves out playing on the tree trunks, in the earth shelters and living willow domes in the nature and adventure playground nearby.

You could also put in a counter where visitors can buy homemade produce (vinegar, herb oil, spirits, juice, jam etc.). When designing a rest area you should not forget to plant plenty of trees to provide shade. Pergolas can also work well here. If there are plenty of sweet-smelling flowering plants nearby, then the trip will become an unforgettable day out and the visitors will happily make the journey to the pick-your-own. Of course there does not have to be a rest area if you do not want your visitors to stay for long. In any case, it is a good idea to provide places to sit down, e.g. a 'rustic' bench made of logs, because some people are not so good on their feet.

Before you begin work on the beds, you should find out what kind of herbs, fruits and vegetables are generally grown in the area, so that you can find a niche for your own crops, which ensures you will have a commercial advantage. In my experience, unusual varieties are always in demand, whether they are fruit, vegetables, salad or herbs. I get requests for my new variety of purple potato from all over Austria. By cultivating and propagating these unusual varieties, you will be making a substantial contribution towards maintaining the diversity of cultivated plants.

Waterscapes

The use and management of water is one of the most important areas of Holzer permaculture. Water is life, so it is of great importance to deal with this resource carefully. Unfortunately, these days very little thought is given to water

Waterscape on the Krameterhof.

conservation and using water carefully in agriculture. In the past, we tended to treat water with great respect either for financial reasons or just out of instinct. Many 'modern' agricultural techniques have led to the problems we are dealing with today. People are abandoning the cultivation of small-scale plots of land, grubbing out hedges and creating the man-made steppes of the monoculture landscape.

These areas do not just lose humus to strong winds; they also dry out much more quickly. In most of them the water table sinks constantly and any water that is left can only be accessed with the frequent use of a pump. The water is usually so contaminated by the large amounts of fertiliser and pesticides that it is no longer suitable for drinking. Trees next to streams are often felled to make room for just a few more square metres of cultivated land. This means that there is little left to stop fertilisers and pesticides from reaching the watercourse. Drainage and straightening streams also have a substantial impact on nature. The straightening and channelling of streams has a large influence on natural flora and fauna. It is my opinion that these techniques frequently have the opposite effect to that which was originally intended. Nature will not be boxed in or subjugated. A stream that can no longer flood the land at its headwater during heavy rains will carry the water further downstream where much greater flooding will inevitably occur. In the future, we will need to take a step back and reconsider our approach, because these problems will not be solved if we continue to act as we have done so far.

Permaculture principles lead us to use water as wisely and carefully as possible. In a permaculture landscape every resource and advantage is made use of exactly where it is. This is why I often consider drainage to be a bad idea. Wherever water is found is where it should be made use of. If I want to make money out of wetlands I do not drain them, instead I cultivate plants like orchids or a variety of aquatic and marsh plants that prefer damp or wet conditions. Taking the next step and putting in a water garden or pond will require only a minimal amount of energy if they are placed in areas that are naturally damp or wet. By doing things this way and paying attention to the natural conditions of the land I have always had excellent results.

I also try to keep water on my land for as long as possible and make as much use of it as I can on the Krameterhof. From the upper boundary of the farm (1,500m above sea level) to the valley (1,100m) the water is used for a number of different purposes: the springs are used for drinking water and are also used to feed the ponds and water gardens. These ponds are placed throughout the entire farm and most of them are connected. In total, a network of 60 ponds, water gardens, wetlands and ditches covers the Krameterhof. They allow me to breed fish, crayfish and mussels as well as cultivate many different aquatic plants. Ducks and geese can also be kept on the ponds quite happily.

Areas of water also provide a number of advantages that are not so obvious. Wetlands, water gardens and ponds provide a habitat for innumerable useful creatures such as snakes and amphibians. They are some of my hardest 'workers' and the ponds ensure that there are more than enough of them. These workers play an important part in the regulation of so-called 'pests'. The toad (*Bufo bufo*) is particularly useful, because its favoured prey is the dreaded Spanish slug (*Arion vulgaris*). Another advantage of waterscapes is the positive effect they have on crops. Large areas of water help to balance out temperature fluctuations on the hills next to them by reflecting sunlight and releasing stored heat. They

Tadpoles in a shallow area of water.

increase the moisture levels in the soil and create useful microclimates through evaporation.

We even use water power on the Krameterhof: the farm reaches from 1,100m to 1,500m above sea level, so I can use the height difference to create energy in an environmentally friendly way. I built two small generators using old Pelton wheels, which are fed from the ponds through a penstock. The height difference of, in this case, around 100 metres (10 bars) allows me to supply the whole farm with electricity. I have already described how to construct a generator of this kind in detail, with all the bureaucratic obstacles it entails, in my book *The Rebel Farmer*.

I also make use of other more traditional ways of utilising water power. I have two mills which are driven by water wheels. There is also a hydraulic ram in use on the farm. It is powered using water from the pond and, using the pressure difference, it pumps drinking water without needing any extra energy. The drive water constantly stretches a membrane inside the apparatus as it pulses. This way drinking water is pumped up through a delivery pipe to the mountain pasture. It provides up to 15 bars of pressure. The flow rate ratio is 1:10. To pump one litre of drinking water I need 10 litres of drive water. Apart from creating energy, these systems also have the advantage that they all release cool, oxygen-rich water back into my fish ponds. This allows me to keep trout in the lower, warmer ponds without difficulty.

The subject of aquaculture is so broad that it could fill an entire book on its own. This means that I will only be able to cover some of the basic principles in the following section.

Building Water Gardens and Ponds

Before you build a pond or water garden it is important to have a clear idea of what it will be used for. For each function you will need to take different things into consideration. A pond for fish or crayfish has very different requirements from a water garden for aquatic plants or a natural pool for bathing. Of course, it is possible to combine all of these functions, but you will have to take this into account from the outset. As it would go beyond the scope of this book to discuss these points in greater detail, I will explain my methods using the example of a pond and its many possible uses on the Krameterhof.

The most reliable example of a functioning, ecologically valuable and visually pleasing pond is found in nature. So if you decide to build a pond, you should take a good look at a natural body of water first. This is the only way that you can get a good grasp of the basics and you will also find a constant source of ideas and possible designs. A pond can only fulfil its purpose once it has developed into a functioning ecosystem that can be used simply and with little expenditure of energy.

My experience with building ponds stretches back for 40 years. I made my first small pond using just my bare hands. In time, I learned from these experiences and went on to build larger and larger ponds. Do not forget that you will still need approval from the water regulatory authority, which usually involves having a stability survey carried out. I think that it is important to start small so that you can gain plenty of experience for yourself. However, if you want to build a large pond straight away you will need to consult an experienced professional.

I begin by surveying the site for the pond. I take note of the soil conditions, topography (position in relation to the terrain) and any water that is present. The topography is important for the stability of the pond. Building the pond in the proper way will prevent it from leaking or sliding. It is vital to get a good idea of the soil conditions and to find out if there are any parts of the land that are naturally wet or areas where a landslide has already occurred. Naturally occurring water is a real advantage when building a pond. Of course, water can also be fed into the pond, but this requires considerably more work.

If there are no springs and the ground water cannot be used, you can always make a pond that is fed by rainwater. However, these ponds are generally more suitable for aquatic plants. Fish and crayfish require a constant exchange of water, because they need it to be fresh and rich in oxygen.

The shape of the pond must look as natural as possible. It is important to have a well-structured pond with both deep and shallow areas. This makes it possible to have a functioning ecosystem, as different plants and animals also need different habitats. The better the pond is structured, the greater the number of functions it can fulfil. Shallow areas create a habitat for a large

This pond at 1,500m above sea level on the Krameterhof is used for growing aquatic plants and keeping fish and crayfish, as well as for bathing.

variety of plants and animals and allow many different kinds of fish to reproduce naturally. This also rounds the pond out nicely and makes it easier to climb in and out for bathing.

Deeper areas of water are also required so that the fish can hibernate and in order to prevent aquatic plants from becoming overgrown. Most of the invasive plants grow to a depth of around two metres. If I want to prevent the pond from becoming overgrown, I make part of the pond about three metres deep with a sharp drop from the shallow area. This creates a barrier the plants will not be able to penetrate. This is only one example of how a well thought-out plan can save a great deal of time and effort later on. Deeper areas of water are also important for balancing out the temperature in the pond. They make it possible for the fish to choose between warm and cold water according to their needs.

Once I have a good idea of the local conditions and have decided on the shape and size of my pond, the work can begin. First of all, the rough shape of the pond is dug out. The size and type of machinery that I use depends on the terrain and the size of the pond. Small wetlands can be made by hand, whereas larger projects will require a mechanical digger. When building the walls of the pond it is important to separate the coarse from the fine material. To do this the material is heaped into a tall pile. The coarse material will roll away to the sides and the fine material will stay in the middle. The pond walls are now made from 30 to 50cm thick layers of fine material and tamped down. With larger ponds

Finished pond on the Krameterhof.

the layers should be compacted with a digger. The coarse material can be used later on for stabilising the walls and banks.

Islands and small biotopes can also be made in the pond. Once I am happy with the shape of the pond, it must be sealed. Water is fed into the pond until the digger is standing in 30 to 40cm of water. Then a small excavator bucket vibrates the subsoil from half a metre to a metre deep. The depth depends on the local soil conditions. The excavator bucket is inserted into the earth and vibrated, which makes the fine material sink and makes the base of the pond watertight. The effect is similar to vibrating concrete. To get the most use out of the pond later on, a standpipe is put in at the deepest point of the pond during construction. The water level can then be regulated simply by adjusting the height of the pipe. This way the pond can be drained at any time to 'harvest' the fish and plants. In case of heavy rainfall, I always put in an emergency overflow pipe. The pipe can take in large amounts of water and allow it to drain away safely.

This is the method I use on the Krameterhof. On loam soils the material does not need to be separated – except for the humus layer which needs to

Stones used as a crossing and to store heat in a water garden.

be carefully removed, separated and put back again afterwards when carrying out any kind of work. Sealing the pond is also much easier with loam soils. I have already described the particulars of building pond walls in the section 'Experiences with Different Types of Soil'.

Design Ideas

Once the pond is finished, I begin to shape the banks. To do this I use stones and tree stumps. Stones rising out of the water warm up in the sun very quickly, which raises the water temperature. In winter, this reduces the length of time the surface of the water is frozen and also reduces any danger of the fish not getting enough oxygen. Water gardens containing fish and plants that prefer warm conditions profit from this most of all. At this stage, I can let my imagination run wild and make my vision a reality. From picturesque, gnarled tree trunks, stone steps and crossings to bridges, everything can be done at very little cost. The excavator just has to lay stones in the shallows or hammer logs into the base of the pond as supports for a bridge.

My method for building ponds is completely at odds with conventional methods where the pond is made watertight with a liner. I am convinced that a wildlife pond should not have a pond liner as it prevents natural pond life from developing. The 'vibration method' can be used on almost every kind of subsoil and is usually cheaper than building a pond with a liner, because the cost of hiring the excavator is generally much cheaper than the cost of the liner and underlay required. I would also not be able to work with this kind of pond any further, because the sheeting would be too fragile and easily damaged. Growing aquatic plants as well as changing the shape of the pond or working on it using mechanised equipment would be out of the question.

Koi carp in a pond at 1,500m above sea level.

Possible Uses

A pond for aquatic plants, fish or crayfish does not have to be square or even made of concrete to be used efficiently. The right way to create high quality produce is nature's way. For example, I have successfully kept brown trout (*Salmo trutta*), arctic char (*Salvelinus alpinus*), carp (*Cyprinus carpio*), tench (*Tinca tinca*), pike (*Esox lucius*), wels catfish (*Silurus glanis*), zander (*Sander lucioperca*), koi carp, forage fish like roach (*Rutilus rutilus*), rudd (*Scardinius erythrophthalmus*) minnows (*Phoxinus phoxinus*) and also European crayfish (*Astacus astacus*) and swan mussels (*Anodonta cygnea*) in my ponds and water gardens for decades.

Yellow irises, white water lilies and bulrushes in a water garden.

I have also cultivated aquatic plants up to a height of 1,500m above sea level such as: many different varieties of white water lily (*Nymphaea alba*), yellow water lily (*Nuphar luteum*), bulrushes (*Typha latifolia*), sweet flag (*Acorus calamus*), fringed water lilies (*Nymphoides peltata*), water plantain (*Alisma plantago-aquatica*), water soldier (*Stratiotes aloides*), mare's tail (*Hippuris vulgaris*), arrowhead (*Sagittaria sagittifolia*), yellow irises (*Iris sp.*) and many others.

I grow the plants in the shallow areas of my ponds and in specially made ditches. As these ditches are shallow, the temperature of the water is higher, providing aquatic plants, which prefer warmer temperatures with the perfect conditions to grow. The plants can be easily harvested from paths running next to them. As I grow unusually hardy varieties without fertilisers or any additional care, I can also replant them in very unfavourable locations and they will thrive where other plants would be unlikely to. When building new ponds and wetlands I can make good use of these tough and hardy plants.

However, ponds and water gardens are more than just a way of making money, they also delight the soul. Water is life. Anyone who has listened to the frogs croaking in the evening or just sat by the side of the water quietly for a while will know why.

2 Alternative Agriculture

Basic Ideas

In my experience, a great deal of the problems with conventional agriculture are caused by the dependence of many farmers on subsidies, the government and cooperatives. They do this, because they believe it will guide them in the right direction. However, this is usually a terrible mistake, as these areas are frequently strongly influenced by agribusiness, agrochemicals and the lobbyists working for them. The training given in many agricultural schools, colleges and university courses often seems to be one-sided and focused on fulfilling the requirements, wishes and demands of agricultural lobbyists. Scientific research projects are supported by wealthy companies and guided in a direction that suits them. There seems to be very little funding available for research into the principles and practice of sustainable farming, permaculture and the interactions between different plant communities. This is because these principles would not help to increase the use of pesticides, chemical fertilisers and specialised technology, in fact, they would minimise it.

There are more and more people following the path of conflict instead of accepting nature in all its diversity. The belief that nature can be improved upon and so-called 'pests' should be fought against is a mistake. When an imbalance emerges, we have to establish the source of the problem and not just treat the symptoms.

In the majority of cases, the specialisation and modernisation of farming practices has not given farmers the advantages they hoped for. It has only forced farmers – who were still well respected when I was a child – to rely on subsidiary income to keep themselves in business. Many farmers now grow large quantities of a far narrower range of crops. To do this they need to invest in expensive buildings for animals, and crops, and specialist equipment and machinery, which can usually only be used for one purpose. This specialisation makes it difficult to react to the market and unforeseeable changes. The produce is usually marketed by a bulk buyer who decides the price and conditions of purchase. This results in a one-way dependence. Changing to another way of farming is usually difficult for these farmers, because they generally have a large number of commitments like subsidy contracts and agricultural credit. Even investments, which have already been made, make farmers reluctant to change their methods, because there would suddenly be no use for the new building that can hold a hundred fattening pigs. So they go on as they did before and they remain at the mercy of the market and subsidy cutbacks. When there is a shortage of money, many farmers try to compensate for it by increasing production. This is completely the wrong approach! One of the biggest problems is that so many farmers are fixated on subsidies. Contrary to all of the promises, it should be clear to anyone that the subsidy system, as it exists today, will not last. Subsidies should never be the main source of income for a business!

The beneficiaries of industrialised agriculture are cooperatives, companies and lobbyists for agrochemicals and agribusiness, but not farmers. We are now familiar with the full consequences of this: intensive livestock farming, groundwater pollution and contaminated food to name but a few. We desperately need to change the way we think.

The ones who suffer from this development most are the farmers' families, who often cannot cope with the strain any more – and naturally the livestock suffer as well, as they are forced to eke out a pitiful existence.

Fortunately, a brave few leave the path of conventional agriculture and dare to follow their own ideas and vision. This is why it is so important to have courage and determination. If you are used to doing all of your work by following 'patterns' and 'recipes' you will find that this path towards a new and genuine independence takes some getting used to. You will have to make all of your decisions for yourself. What your neighbours are doing is not necessarily right for you – quite the contrary: what everyone around you is currently growing or breeding is already available in abundance and is therefore no longer interesting as a product. This method also requires courage, but it is worth it if you proceed cautiously but with determination. Ecological farming can also make sense economically, as the Krameterhof shows. Before the business began to offer training and excursions, it was a full-time farm. However, I would not advise anyone to try the same strategy, because it is your own strengths and interests that help a farm to grow and nothing else. There are plenty of niches in the market. You only need to use your intuition and take a good look around you. It is important to remain flexible and not to invest money in rigid farming methods that will become unprofitable the second the market changes. My experiences and ideas along with the old farming methods that I want to bring back should encourage you to think and act independently again. The goal is to find alternative farming methods for your farm that are based on natural cycles and allow you to live in peace and harmony with nature.

The basis for farming is the fertility of the soil. In the next section I will discuss this topic in greater detail.

Soil Fertility

A healthy soil that is rich in microorganisms is a fertile soil. This is a fact that a farmer should never forget, because it is the main requirement for successful farming. If this is always ensured then the farm will always remain flexible. You must pay attention to natural processes and try to make use of them. If you treat the land with care nature will work for you. I want to stress that it is vital for

A mixed crop of soil-improving plants.

our attitudes to change in this area. Enough damage has already been caused by conventional monoculture farming methods and the excessive amounts of pesticides and fertilisers required to maintain them. The soil should not be seen purely as a production plant, it is a diverse and sensitive ecosystem. Innumerable creatures play a part in maintaining this system. It is only with their help that the soil will remain fertile and of use to us.

Every plant has its own requirements and affects its surroundings and the soil in a different way. If there is only one kind of plant growing in a certain area, then the demands on the soil are unbalanced. If the crop is harvested completely, the nutrient content of the soil will become lower and lower until it is completely exhausted and only large amounts of fertiliser will briefly make it suitable for growing again. Then the topsoil will be left bare through the harsh winter, which makes it even more difficult for the microorganisms in the soil to thrive, assuming they have not already been killed off by the chemicals. If the exhausted soil is to be regenerated, we have to look after the soil life first. The creatures living in the soil – earthworms, bacteria and fungi among others – are the key to healthy soil. In order to provide them with the right environment, it is important to avoid using pesticides and chemical fertilisers. The common practice of deep ploughing in the autumn causes the soil to freeze and they both in turn destroy not only the soil life, but also the natural layering and the build-up of humus. If you leave these areas fallow for a while, they usually regenerate on their own. This process of regeneration is self-supporting. By growing plants that improve the soil, nature can be helped and then nature will start to take care of itself.

Green Manure

The correct plants to select naturally depend on the current state of the soil. I try to bring the nutrient content in overfertilised areas back into balance using very demanding plants. If I want to recultivate contaminated or exhausted soil, I have to take care of the soil life first. In these areas it is particularly important to get a good layer of humus. This is why I try to use as much biomass on the soil as possible. A good mixture of green manure crops is very important, so that the individual plants can propagate themselves. This increases the stability of the system and also its value for the soil, soil life and beneficial insects.

When biomass is left on these areas it benefits the soil and the soil life. The slow decomposition of the green manure crops in autumn and winter builds up a productive layer of soil, which will be well supported by the regenerative power of nature. The biomass and the loosening of the soil (caused by the rootsystems from the plants introduced) leads to a good soil structure, which is the most

Colourful mixture of green manure crops on a terrace.

important requirement for good general plant growth. An additional advantage is that a plot of land worked in this way always provides a large amount of ground cover. This protects the soil from extreme weather (wind, storms, rain, heat and sun) and the plants will retain water and nutrients. The plant cover works like a quilt and protects the soil from frost, so that it freezes much later and the frost does not reach as far into the soil. This means that the soil life can continue working in the topsoil up until late autumn and continue into winter. As I walk around the farm, I regularly check the condition of the soil in different places throughout autumn and winter. If I can dig up earth that has not frozen yet from beneath the snow, it is a sure sign of a job well done.

Green Manure Crops

Above all others, legumes make the greatest contribution to soil improvement. With their varied and distinctive rootsystems (from shallow to deep roots) they can grow in very different areas. The greatest advantage of legumes is that they can fix nitrogen with the help of bacteria and release it into the soil. The bacteria (primarily types of *Rhizobium*) live in close symbiosis with the plant's rootsystem and create root nodules. In the root nodules nitrogen, which is abundant in the air, is fixed and released into the plant's nutrient cycle. In return, the bacteria receive carbohydrates from the plant, which helps them to grow. This symbiotic relationship offers the bacteria nothing but advantages. Once the plant dies it rots down into nutrient-rich humus. After this, there are far more Rhizobia in the soil than there were originally, so it is not only the original symbiotic partners that profit from the action of legumes, but also the entire area.

The best-known members of this family of plants belong to the sub-family *Faboideae*. This plant family is very large and is found all over the world. Thanks

Lupins improving the soil.

to their symbiotic relationship with bacteria they grow well on dry and nitrogen-poor soil. Some of the members of this family are: peas, beans, clover and lupins. However, legumes are not the only good green manure crops, a number of varieties of cabbage, oilseed rape, turnip, sunflower and buckwheat are also excellent at improving the soil, because they grow large amounts of leaves and fruit within a short period of time.

The root nodules on this lupin can clearly be seen.

My Method

In autumn I usually leave green manure crops standing. Nature does all the work for me: the first heavy snow pushes down the plants and they begin to decompose. In my opinion, this is the best and most successful way to improve the soil. The plants rot down slowly and the biomass does not compact so easily, which would usually be the case with material that is cut down. The plants grown in mixed culture vary so much in height and structure that the area looks quite 'overgrown' by autumn, which means that the biomass does not collapse, but instead gets roughly packed together. On land managed in this way the air circulation is always good and the conditions for soil regeneration are optimal.

A plot of land, which is managed using this method, provides enormous advantages and can also give an adequate yield. I can alter the land as often as I like, growing different main crops and managing it using different systems (orchard, paddock etc.) – or work it with or without the use of animals. The area can also be used for gathering seeds or to grow honey plants. The proportion of honey plants, medicinal herbs or culinary herbs – in other words the plants

My method: in autumn I leave green manure crops standing.

grown in addition to the main crop – can be varied to suit different requirements and the intended purpose of the land. Ancient cereals are also well suited to this. The more diverse the plants are, the more stable the system is and the more useful it can be.

Areas where green manure crops are cultivated have the advantage that the composition of the plants can have a significant effect on animals and insects. I often sow sunflowers and hemp, because these plants are an excellent source of food for birds. To encourage useful nectar- and pollen-collecting insects (bumblebees, bees, lacewings, hoverflies etc.) various local wildflowers such as cornflowers (*Centaurea cyanus*), yarrow (*Achillea millefolium*), calendula (*Calendula officinalis*), golden marguerites (*Anthemis tinctoria*), scented mayweed (*Matricaria chamomilla*), spreading bellflowers (*Campanula patula*) and comfrey (*Symphytum officinale*) make a suitable addition.

Leaving green manure crops standing means that I save myself more than just the work of cutting the plants down: this way the plants can ripen, bloom and produce seeds. I no longer have to reseed the area. Many seeds are eaten by the birds, stratified in their stomachs (the layer which prevents germination is broken down) and are distributed in other areas. If I had to continually sow and plant my 45 hectares of land, this in addition to all the other things I have to do would take up far too much of my time. It would also be too expensive to have to constantly buy more seeds for sowing large areas of land – wildflowers are incredibly expensive!

The addition of various flowering plants will help green manure crops to provide a better habitat for useful animals and insects.

Lupins improving
slope stability and
soil conditions.

On steep slopes and embankments I make sure that there is a large number of very deep-rooted plants like lupins and sweet clover in my seed mixtures. These plants do not only improve the soil, but they also stabilise the slope with their deep rootsystems (strong main and taproots). Their strong growth and well-established roots also improve the capacity of the soil to retain water. When you compare a slope of this kind with a meadow under similar conditions, the difference becomes even clearer. On the slope the lupin and sweet clover roots are metres long, whereas in the meadow there are mainly grasses with roots measuring only centimetres. In my polycultures, I have the soil and plants 'working' to a depth of metres and not centimetres. On steep slopes and embankments this effect is of greater importance, because without it heavy rainfall would easily lead to high surface runoff, the erosion of humus and landslides.

A mixture of sweet clover, lucerne, vetch, peas, lupins, sunflowers and different tubers like Jerusalem artichokes and turnips is well suited to this. This mixture of plants and the slow process of decomposition will activate the soil life very quickly. With the help of this low-work method of not cutting green manure crops, I managed to improve the poor and dry soil on my slopes to the extent that I could grow demanding fruit trees after only two to three years. This is how I changed the slopes on the Krameterhof into the lush orchards that they are today.

Mistakes

The widespread use of flail mowers, which are very common in Burgenland and Steiermark, stands in opposition to the principle behind environmentally friendly green manure crops. All of the vegetation is shredded to a fine material. The fact that this also kills everything from the tiniest creatures up to creatures the size of a ladybird is not considered. The shredded material dries out quickly and is mostly blown away by the wind or washed away by the rain. The result: the soil is left bare and defenceless against erosion. Erosion dries out the soil and deep cracks form (especially in loam soils). Fine particles of soil are carried away by the wind and this has a negative effect on the soil life. Then the soil's ability to retain water suffers. Heavy rainfall leads to flooding and landslides. The water table sinks as a result of the lack of retention, which causes springs and wells to dry up. Last of all, the soil loses its natural ability to regenerate. Mankind's use of chemicals has an even greater effect. Luckily, we are not powerless in the face of these developments! Nature's regenerative processes can be properly supported with the help of green manure crops grown in mixed culture.

Fiddleneck (*Phacelia tanacetifolia*)

Plant List

I have put together the following list to give you a quick overview of the best green manure crops.

Gold-of-pleasure (*Camelina sativa*)

Common Name (Family)	Botanical Name	Notes
Legumes	*Fabaceae*	
Yellow lupin Narrow-leaved lupin White lupin	*Lupinus luteus* *Lupinus angustifolius* *Lupinus albus*	annual, good for slope stabilisation and building humus, grow well on sandy and acid soils, forage plants, honey and insect plants
Garden pea	*Pisum sativum*	annual, good forage plant
Grass pea	*Lathyrus sativus*	annual, undemanding
Fodder vetch	*Vicia villosa*	perennial, light soil, good nectar plant
Narrow-leaved vetch	*Vicia sativa*	annual or biennial, undemanding
Faba bean Field bean Broad bean	*Vicia faba* *Vicia faba minor* *Vicia faba major*	annual, loose soil, good forage plants (rich in protein)
Yellow sweet clover White sweet clover	*Melilotus officinalis* *Melilotus albus*	biennial, also grow on dry soil, permanent green cover, good as catch crops
Red clover	*Trifolium pratense*	biennial or perennial, permanent crop, forage plant
Subterranean clover	*Trifolium subterraneum*	annual, acid soil, permanent crop, good as a catch crop
Alsike clover	*Trifolium hybridum*	perennial, forage plant
White clover	*Trifolium repens*	perennial, permanent crop, forage and meadow plant, all soils, good as a catch crop
Crimson clover	*Trifolium incarnatum*	annual or biennial, permanent green cover
Persian clover	*Trifolium resupinatum*	annual, frost hardy, also grows on poor soil
Egyptian clover	*Trifolium alexandrinum*	annual, frost hardy
Kidney vetch	*Anthyllis vulneraria*	biennial, also grows on poor soil, good as a catch crop
Birdsfoot trefoil	*Lotus corniculatus*	perennial, hardy, good as a catch crop, permanent green cover
Lucerne	*Medicago sativa*	perennial, permanent green cover, also grows on dry soil, good forage plant, good for slope stabilisation
Black medick	*Medicago lupulina*	perennial, undemanding, good as a catch crop
Sainfoin	*Onobrychis viciifolia*	perennial, grows on chalky soil, pioneer plant, permanent green cover, honey plant, good forage plant
Serradella	*Ornithopus sativus*	annual or biennial, acid soil, prefers sandy soil, forage plant

Common Name (Family)	Botanical Name	Notes
Crucifers	***Brassicaceae***	
Oilseed rape	*Brassica napus*	annual (summer variety), biennial (winter variety), undemanding, good catch crop
Turnips	*Brassica rapa*	summer and winter crop, similar to oilseed rape, undemanding
Radishes	*Raphanus sativus*	annual, forage plant
White mustard	*Sinapis alba*	annual, undemanding, pioneer plant, frost hardy
Marrowstem kale	*Brassica oleracea var. medullosa*	annual, very good forage plant
Grasses	***Poaceae***	
Rye Wild rye	*Secale cereale* *Secale multicaule*	perennial, frost hardy, undemanding, very good forage plants, good yield
Sorghum Millet	*Sorghum dochna* *Panicum miliaceum*	annual, prefer sunny areas
Others		
Buckwheat	*Fagopyrum esculentum*	annual, good honey plant
Fiddleneck	*Phacelia tanacetifolia*	annual, undemanding, grows on all soils, good honey plant
Sunflowers	*Helianthus annuus*	annual, good honey plant, seeds can be left as bird food
Jerusalem artichokes	*Helianthus tuberosus*	frost hardy, undemanding, needs well-drained soil, seeds well, tubers will produce new plants, very good forage plant
Flax	*Linum sp.*	annual, oil and fibre crop
Gold-of-pleasure	*Camelina sativa*	annual, undemanding and fast growing, can grow in poor soils (sandy), high drought resistance, fairly disease and pest resistant, oil crop
Salad burnet	*Sanguisorba minor*	perennial, undemanding, also grows on chalky soils, year-round plant cover
Mallow	*Malva silvestris*	perennial, year-round plant cover, medicinal plant, good honey plant

Ways to Regulate Problem Plants

It is important to always bear in mind that when we work a plot of land we are changing the natural balance to a greater or lesser degree to make it benefit us. Cultivated plants are usually not as well adapted and strong as wild plants, so we try to help them grow by doing things such as tilling the soil. Sometimes it is also necessary to regulate competition. It is important to bear in mind that everything in nature has a reason. We must try to understand the natural processes and influence them in our favour. Just fighting the symptoms of a problem will not work – especially seeing as we have caused most of these problems ourselves. It is not just nature and its 'catastrophes' that are to blame when large areas of forest are suddenly brought down during a storm, the managers in charge of the unstable monoculture systems are also to blame. Bark beetle infestations are a result of these unnatural farming methods as well. Short-sighted thinking creates these problems in the first place. We have to admit these mistakes to ourselves. Anyone who tries to look at nature with open eyes will soon recognise that there is a reason for everything and a solution to every problem.

Controlling and regulating things on a small scale is quite simple. Almost any desired effect can be achieved with manual labour, which leads many people to take their obsession with order too far. They often do not consider the consequences of their actions.

At this point I would like to make these consequences clear with the following example: I have a small garden and want to make it neat and tidy. I remove all of the weeds in my vegetable patches. I keep the lawn short and the ground under my fruit trees 'neat'. What will I achieve by doing this? The answer: a tidy, in other words, 'artificial' garden. There is nothing left to stop the vegetable patches and fruit trees drying out because of the lack of ground cover, so I will have to water them more.

The humus production on bare soil is much worse and frequent watering flushes out the nutrients, which means that sooner or later I will have to use fertiliser. Chemical fertilisers are bad for the soil life, fewer creatures living in the soil makes the humus production worse and the vicious cycle continues. A 'tidy' garden provides useful animals and insects with little refuge and no habitat, which means that there are no natural defences against pests. The list goes on. This just illustrates the relationship between action and reaction in nature. If I manage my land with an understanding of nature, I can achieve great results with much less work. My methods for gardens are outlined in the 'Gardens' section.

On agricultural land things are no different, the only difference is that the work is carried out on a larger scale and must be better thought out. However, what generally applies to small plots of land also applies to larger ones. Managing crops or pastures incorrectly and in an unbalanced way often leads to individual plants getting out of control and driving out the cultivated plants that you want to grow. I will use broad-leaved dock (*Rumex obtusifolius*), stinging nettles (*Urtica dioica*) and orache (*Atriplex patula*) as an example. These plants indicate a high level of nitrogen in the soil. Overfertilising or overly intensive pasture management usually causes this imbalance. To deal with this I have to try to restore the balance and manage the land differently. It makes no sense just to treat the symptoms.

Problems with rapidly growing plants frequently occur in fallow areas or on land that has changed from being managed using conventional methods to being managed using natural ones. These areas often used to be intensively fertilised and provide ideal conditions for these plants and suddenly not using pesticides any more lets them come back up. Some farmers then begin to doubt themselves and forget that this change is the right decision. Mistakes that were made years or even decades ago cannot be rectified in such a short period of time. Nature takes a while to recover.

With my farming methods – namely keeping livestock on the same land as my crops – these plants do not pose a real problem. Moving the paddocks ensures that the land is never overgrazed. It has time to recover while not in use or can be used to grow crops. The danger of livestock diseases is almost entirely prevented by the paddocks being moved and their varied diet. If a particular variety of plant still appears in numbers that are too great in one area, there are still possible solutions: I put my pigs out to graze in this area and support the process by sowing peas, beans or sweetcorn between the unwanted plants. This makes the pigs concentrate on these areas. The plants and roots will be partly eaten or the digging will bring them to the surface where they will dry out. After this is done, the pigs are moved on to the next paddock and I introduce demanding plants, especially tubers like Jerusalem artichokes (*Helianthus tuberosus*), but also sunflowers (*Helianthus annuus*) and hemp (*Cannabis sativa*). They absorb all of the excess nutrients and make the conditions worse for the 'weeds'. They also grow tall and quickly on the nutrient-rich soils, so they will overshadow and kill off any remaining problem plants. The cultivated plants provide livestock with a valuable source of food. Jerusalem artichokes, for example, are perennial and give a high yield of tubers. The plants can be eaten by the pigs or, if necessary, removed again. This method helps me to rid an area of unwanted plant growth, balance out the nutrient conditions and still get a good yield.

Another possible way to regulate problem plants is covering and mulching. With these methods we cannot only regulate harmless wild plants, but also invasive plants like sorrel. To cover an area, I use cardboard, jute sacks and other biodegradable materials (i.e. only natural materials). To weigh this down I pile soil and mulch on top. The material covering the plants should not be airtight of course; otherwise everything underneath it will die. The problem plants will not receive any more light under this layer; they will die off and provide the soil life with nourishment. Immediately after I have covered an area I sow the mulch with seeds. For this I use the demanding plants I mentioned previously.

Tubers on a
Jerusalem artichoke
(*Helianthus
tuberosus*)

All varieties of turnip are well suited to this method. The plants will develop well, because the mulch soon provides them with high-quality humus. They overshadow the system and balance out the nutrients within the soil. As broad-leaved dock requires light to germinate, a permanent plant cover usually prevents it from reappearing. However, the broad-leaved dock is very hardy; its seeds can survive in the soil for years and can grow back from its roots. For these reasons it is usually necessary to repeat the covering process. This should be done early in the year, because this will prevent the unwanted plants from seeding. The covered area will also be sown with crops, which will need time to develop. Careful observation will help you to quickly recognise when it is a good idea to intervene. If a certain variety of plant is becoming over prevalent and emergency measures are needed, it will already be too late for the methods I have described and controlling the problem will become more and more difficult.

You should never forget that every creature has its purpose in the cycle of nature and can also be very important to humans. Cornflowers (*Centaurea cyanus*), to name an example, are now relatively rare, because they have been banished from cereal fields as a so-called weed. The fact that they are not only pleasing to the eye, but also valuable medicinal plants generally goes unnoticed. The stinging nettle is also a valuable plant. It performs a number of purposes as a culinary and medicinal plant, for liquid fertiliser and for mulch. It is also indispensable as a food source for caterpillars. In nature there is nothing bad, and there is a solution to every problem. You only have to look for it. Every plant has a natural rival. If you leave them enough freedom within the system they will work for you. For example, the green dock beetle (*Gastroidea viridula*) can become a useful worker. If you give them the opportunity to grow to reasonable numbers, they can help you to keep the system in check. Balancing out the soil conditions is the main priority. A single species can only become prevalent if there is an imbalance.

REGULATING PROBLEM PLANTS

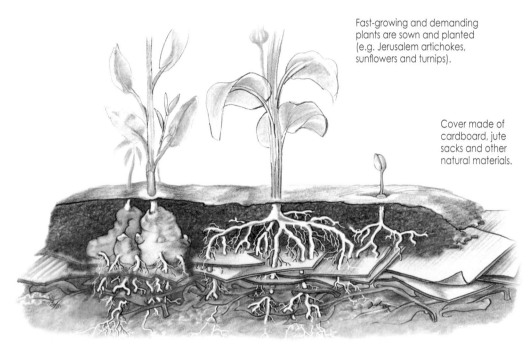

Fast-growing and demanding plants are sown and planted (e.g. Jerusalem artichokes, sunflowers and turnips).

Cover made of cardboard, jute sacks and other natural materials.

The roots grow through the cover and absorb the excess nutrients available after decomposition.

Invasive plants (e.g. sorrel) are covered. As they were not cut first, they will not compact. There is enough oxygen left for them to rot down quickly.

Old Varieties and Plant Diversity

My experiences all show that old varieties of cereal, vegetable or potato – of all cultivated plants in fact – make the best crops. They have a better flavour, are generally less demanding and deal far better with local conditions than new, overbred varieties. In earlier times, every country, in fact every region had their own cereal varieties. This also went for vegetables, potatoes and herbs. There were numerous local varieties which were grown and propagated on only one farm. These varieties did not usually have names of their own. Through a process of selection, they were bred over generations to become the plants best suited to the local conditions. Therefore these varieties grow best in the areas in which they were bred. This is why it is important to ask around in your area to find out what local varieties there are and then to propagate them.

Hybrid seeds (F1 varieties) are entirely unsuited to permaculture systems. These seeds have been bred so that most of them no longer have the ability to reproduce. They do not breed true (in other words they cannot pass on the characteristics and qualities of their variety every time) and must always be bought new again – much to the satisfaction of seed companies. They come from homozygous inbred lines, are frequently sterile and are not suited to varying local conditions. They give good yields, but are more susceptible to disease, because they are not suited to local conditions. All of this coupled with the crops being grown as a monoculture leads to the population of single species increasing to large numbers and later leads to the use of pesticides.

Genetically modified seeds are even more questionable! In my opinion, it is a crime against nature to alter the genetic make-up of a living thing in this way. Anyone who treats nature with consideration and understanding will feel the same. The results and repercussions of using genetic modification in plant cultivation and animal breeding cannot yet be foreseen and it seems that its use can barely be regulated or kept in check any more. As a result of specialisation and modernisation, we have almost entirely lost the plant diversity we once had and, with it, many resources in farming and gardening. The fact that in the EU the propagation and marketing of seeds has been taken away from farmers and has been left to agricultural companies is particularly questionable. Once I became aware of this, I made a great effort to propagate the seeds of every conceivable cultivated plant suited to the Krameterhof. The plant diversity on the Krameterhof provides me with a kind of living gene bank. Visitors to the Krameterhof are allowed to harvest seeds in small amounts for their own use during tours of the farm.

I have established through my experience that the seeds of the strongest plants, which grow on the worst soils and under the most extreme conditions (high altitudes, frost etc.), are the most suitable for propagation, because they have positive energy and establish themselves well. In academic journals the opposite is often to be found. They claim that the seeds from the largest plants, which grow on good soils, should be preferred. As far as I am concerned, these seeds would be the worst choice. Although plants growing on good soils do produce the most seeds, it is my experience that plants bred in this way deteriorate. Seeds collected from the strongest plants on the poorest soils, on the other hand, provide plants that can also deal well with difficult conditions, because they are undemanding and still give satisfactory yields. For me these are the best selection criteria. I also continue to breed hardier and more robust plants which can grow and thrive on their own without the constant support of fertilisers and water. Naturally, I pay attention to the taste of the varieties when selecting seeds. Nutritious and high-quality food develops an intense flavour and contains many valuable substances, to the extent that it works almost like a medicine and protects people and animals from sickness and poor health. Healthy food also develops this flavour. Anyone that has a sense of taste and has been even partially protected from ready meals and fast food can use it to determine the quality of seed-producing plants.

The numerous habitats and microclimates that I create are useful for producing seeds. In these places it is possible to prevent the cross-pollination of related varieties, because they can be isolated spatially from each other. When you are growing different seeds, which can cross-pollinate within a small area, it is important to create multi-layered, structured habitats. If there are hedges and particularly tall raised beds between the individual varieties which are to be preserved and improved, this should protect them from being pollinated, especially if they are wind-pollinated. Varieties that are pollinated by insects should be grown a fair distance from each other, so that they can be bred true to their variety. From experience, there is one thing that you should do to grow seeds successfully. Right at the beginning you should examine the natural reproductive cycles of the plants you have chosen: how are they pollinated, are insects required and, if yes, which? With wind-pollinated varieties you must pay attention to the wind direction, because this is the direction in which the pollen will be carried. If I want to prevent two varieties from cross-pollinating, I have to keep a greater distance between them in this direction. It is simpler to select varieties that flower at different times, because I can then completely rule out unwanted pollination. Nevertheless, many varieties have arisen by chance cross-pollination on the Krameterhof. From these I have propagated the best and tastiest. So there are already a number of varieties of potato, salad and pumpkin that I have had to invent names for.

Cereals

Ancient Siberian grain

Breeding einkorn (*Triticum mono-coccum*), an ancient wheat, and emmer (*Triticum dicoccum*), a very early variety of wheat, is very interesting. Although both varieties only give low yields per plant, it has been proven that they are far richer in protein and contain more minerals than other varieties of wheat. They are extremely nutritious, nourishing and easy to digest. They also cope well with very poor soil conditions. Black emmer is even resistant to UV radiation as a result of its dark colour, which is not the case with any other variety of wheat I have found. Spelt (*Triticum spelta*) is very well known and another old cultivated variety of wheat. Spelt also thrives on poor soil. It has a good flavour and is rich in protein, furthermore it is popularly eaten before it is ripe and can also be used

Ancient Siberian grain in a test area in the Scottish Highlands. It even thrives on this acid heathland.

The undemanding ancient grain also grows on poor soils (here on soil where spruce trees once grew at nearly 1,500 metres above sea level).

as a substitute for coffee. Its positive qualities are so great that it is even used to fill pillows to cure headaches, sleeplessness and tenseness. I have cultivated a number of different varieties of rye such as wild rye (*Secale multicaule*), black oats (*Avena spec.*), naked oats (*Avena nuda*) and barley (*Hordeum vulgare*). A particularly good old local variety of oats we used to grow that does well at high altitudes is *fichtelgebirgshafer*.

On the Krameterhof I also cultivate an ancient grain, which originally comes from Siberia. I was made aware of this ancient Siberian grain in 1957 when I saw an advertisement by a Viennese company in the magazine *Österreichs Weidwerk* that read, 'Ancient Siberian grain – particularly suited for seeding hunting grounds'. I ordered a small amount to try the cereal out. To this day I still cultivate cereal from the one kilogram of seed that I bought back then. It is very similar to our old *brandroggen*; it grows on the poorest of soils and is perennial. This means that if, for example, you sow it in June at 1,400m above sea level, it will ripen the next year in September at the earliest, i.e. the ears will have grown and can be harvested. If it is eaten by cattle or deer or is cut, then it gives a good yield of green material and it compacts well. In this case, ears will grow one year later. Up to 20 ears can grow on one stalk. It was another surprise for me when this cereal ripened in five months at my project in Colombia, which is at sea level and the temperature is around 45 degrees in the shade. This demonstrates the adaptability of this ancient grain.

You must be very careful with fertiliser when growing any undemanding varieties of cereal. If you use too much dung or liquid fertiliser, the grain will grow tall very quickly, then subside and begin to rot in the soil. This manuring increases the nitrogen levels beyond the amount the plants can cope with. It is easier to grow ancient grain in a gravel pit than on fertilised garden soil.

Apart from using them for bread grain and sowing, I also like to use various kinds of cereals as green manure crops and pioneer plants when sowing new plots of land for the first time, because they are very undemanding and grow quickly. I also use them as feed grain. I sow paddocks with a mixture of cereals, legumes and root crops and do not harvest them. The crops provide the best feed for my cattle and pigs in a short period of time and require very little energy.

Growing and Processing *Brandroggen*

Once we used to grow a number of local varieties of *brandroggen*. An example of this is the Lungau *tauernroggen*, which I still grow to this day. As hardly anyone knows about this method any more, I would like to quickly describe how we used to grow *brandroggen*.

In spring, which is usually in May here, it was time for the fields to be burnt, which the children would also help with. This was a task, which involved freeing meadows from branches and plant growth once they were cut. All of the bulky material was raked up, the young shrubs and spruces were chopped up, and everything was put into a number of piles and burnt. Burning made sure that

Rye on a terrace on the Krameterhof.

the meadow did not become overgrown and made it easier to cut later on. It also prevented branches and unsuitable material from getting into the hay. After the piles were burnt, the ashes were raked over the area, then the *brandroggen* was sown. Varieties of rye grown on soil prepared in this way are known as *brandroggen* (literally 'fire rye'), because they are sown after the plant matter is burnt. The cereal grew very well on the meadows prepared in this way. At the end of July or the beginning of August the *brandroggen* was harvested with the rest of the meadow grass, dried and put in the barn. On larger plots of land the *brandroggen* was usually harvested in the second year at the end of August or the beginning of September for bread grain and seed. On the Krameterhof, we still bind the sheaves like we used to and gather them into stooks so that the grain can ripen for two to three weeks in the field.

Then the sheaves were brought into the barn. We put them in troughs and carefully arranged them. We leant the sheaves together in a circle working from the middle outwards so that all the ears pointed to the centre. It was important to make the best use of the available space. When one circle was finished and there was no more space left, we tackled the next circle until all of the grain was in place and the troughs were full. The grain was then stored there until the other autumn tasks were completed and the wood was cut before the snows came. In November it was time to start threshing the grain with a threshing machine, which was powered by an old paraffin or petrol engine. This work was usually carried out when the weather was bad, because nothing could be done outside. It was laborious, dusty and exhausting work. In the threshing machine

On the Krameterhof we still bind freshly cut grain into sheaves then gather them into stooks so that they can ripen in the field.

the ears were chopped up and separated from the straw. Once one variety had been threshed we tidied the threshing machine away to make room for the winnowing machine. This was operated by hand and produced a stream of air to separate the material further. The light material such as the blind ears and pieces of straw were quickly blown away and collected in a large pile that was used later on as bedding for livestock. Often this material was mixed with bran and hot water to form a paste to use as feed. Fine material, like grains that were broken in the threshing machine and particles of sand, fell through the sieve and were collected for chicken feed. Bulky material like stalks or small stones – there were also often deer and hare droppings – was left for the chickens to scratch out the grains. Finally, out of the main chute came the beautifully cleaned seed and milling grain. Everything that came from this process was made use of on the farm, which can only serve as an example to us in today's throw-away culture.

Tips for Polycultures

Growing plants in polycultures is both possible and practical anywhere. As I have mentioned many times before, monoculture farming practices are completely unnatural, they cause many problems and should be stopped. Different crops can grow well as a group, they can be harvested at the same time and processed together (e.g. as high quality feed). Finding the mixtures that work best and give the largest yield in your situation and on your land will take time, especially if you have no experience in this area. As with all experiments, you should always begin on a small scale and observe the crops as they develop before you try the mixture on a larger plot of land. Now I would like to give a few examples of polycultures I have had good results with many times over the years.

When growing cereals, catch crops such as clover, radishes, salad and various medicinal herbs can be sown. These catch crops should be sown once the cereal crops have flowered. While the cereals ripen the catch crops develop very slowly. When the cereals are harvested, the catch crops suddenly receive more light and begin to thrive. Then they can soon be harvested. This method is tried and tested. Catch crops of mixed varieties of clover can be sown earlier. The clover supports the cereal crops while they are growing and helps to prevent rival plants from developing by providing ground cover. When the cereals have been harvested the field can be used as a forage area.

Stubble drilling can also be used. The plants that are suitable for this depend on the location and therefore growing season. Here the cereal harvest takes place in September, so only forage or winter fruit, like winter rape and turnips can be sown afterwards. They are still taking root in autumn, they survive the winter and ripen next summer. Then they are either harvested or left to improve the soil, possibly they could also be tilled in. At lower altitudes where some cereals are threshed in July, a catch crop gives another good yield within the same year. Fast-growing vegetables, turnips and salad are suited to this. After the catch crop is harvested the field can be planted with winter crops in the autumn. Naturally, this kind of intensive use is only possible when none of the crops are grown in a monoculture. In order to provide the crops with the nutrients they need, a balanced mixture of plants which improve the soil and fix nitrogen is required. The polycultures will not exhaust the soil of any one nutrient, because they have different requirements. It is also important not to harvest everything, but instead to leave some of the crops as mulch and a source of nutrients for the plants as green manure. When the second crop itself is used as green manure, it provides an immediate and direct yield, which is normally forgotten: namely seeds for sowing other plots of land.

Further examples are maize, sunflowers and hemp, which can all be grown with peas or beans. The tall-growing plants give the peas and beans something to climb. In turn the peas and beans provide nitrogen and improve the growth of the supporting plants. A culture of Jerusalem artichokes with maize, peas and beans is particularly suitable and they can be harvested together and processed in a forage silo. A catch crop of white clover, black medick and, on wet soil, Alsike clover can also be used.

Peas and maize grown together provide a very good feed combination. They can also be harvested and threshed together. Maize is a very energy-rich feed, but it contains little protein, so adequate amounts of protein-rich feed should be added to it – peas are particularly suitable for this.

Cereals with a catch crop of different varieties of clover.

When growing flax a catch crop of white clover is advisable. If the clover is sown a little later, the flax will have a head start and it will stop the clover from becoming overgrown. When selecting catch crops it is important to pay close attention to the combination of plants, so that the main crop will not be overwhelmed and the plants will not be in competition with each other.

Poppies also make a good addition to black medick and white clover. Buckwheat, which is sown in the spring, also grows very well with white clover as a catch crop. Salad or radishes can also be added. As buckwheat needs a great deal of light it cannot itself be used as a catch crop.

A good mixture for animal feed that we use, is Jerusalem artichokes, varieties of kale and turnips. Jerusalem artichoke tubers survive through the winter and only the leaves and stem freeze. Turnips can withstand frost to a certain extent and the varieties of kale mentioned survive a relatively long time in frosty conditions. This makes it possible to feed livestock naturally for longer. If the snow is not too deep, the animals can even survive the winter on this mixture of feed without requiring any additional feeding.

A polyculture on a terrace on the Krameterhof: a colourful mixture of green manure and arable crops (e.g. tobacco for seeds) thrives. Amongst them grow fruit trees.

You will have to find the best composition of seeds for yourself through experimentation, because this depends a great deal on soil conditions, moisture levels, temperature and factors like wind and frost. The intended purpose of the plot of land also affects the kind of polyculture you will need. If I want to create a pick-your-own area or grow for market, I sow the appropriate kinds of vegetable and select catch crops so that I can always offer a good range. However, if I want to grow feed for livestock that I am keeping in a meadow, clover with vegetables as a catch crop would make an excellent source of food. These are just a few examples to encourage you to start thinking about what kind of mixed culture would be best for your own situation.

In the 'Gardens' section there is a list of plants, which should help you to choose a suitable polyculture for growing in a field. It includes information on the most effective plant communities.

Alpine Plants

Over the years I have had a great deal of experience with cultivating plants from Alpine regions. One of the most important things I have learnt from these experiences continues to shape my work: namely that all events in nature should be observed closely. This is a wonderful and fascinating activity for anyone who has an interest in nature, because you never cease to learn and profit from it. The example of yellow gentians (*Gentiana lutea*) illustrates this clearly:

For a time I tried to cultivate yellow gentians with no success. I tried to grow the plant in sheltered conditions with the help of a variety of different growing instructions – but I was unsuccessful. It was only when I left the seed trays outside the door for disposal and forgot about them (and the seed trays were exposed to the elements for a number of months) that, to my surprise, I suddenly had the result I had given up on. The gentian seeds ripened at an altitude of 2,000m, in other words, in the Alpine region! How did this happen? The explanation is that the weather on high mountains during spring is very changeable. This means that the mornings are frosty, during the day it is dry and warm and later on it rains and snows again. This pattern of weather begins in spring and continues into June, when it gets hot for a short period of time. Therefore gentian seeds get wet, warm up and are frozen many, many times in their natural habitat – like the seeds in the trays I left outside. The seeds germinate in the warmth of summer (July). In the first year the tiny seedlings can hardly be seen. The young plants are, of course, exposed to the same conditions in autumn and winter that the seeds were exposed to the year before. Gentian seeds naturally germinate only under extreme climatic conditions (i.e. frost). So it turned out that I had simply tried too hard with my plants. Nature cannot be 'improved' upon.

If you want to grow yellow gentians you should take into account that they germinate in frosty conditions, so they should be sown in the autumn, winter or spring at the latest, as long as there are still a good number of frosty nights left. If this is not possible, there are ways to simulate these conditions. The seeds can be placed in a plastic bag and mixed with earth from where the parent plant grows. Then some water is added. The bags are then put in a freezer at -10 to -15ºC for a number of weeks, before the seeds are sown. I think it is usually a good idea to take some earth from near the roots of the parent plant when cultivating alpine plants, e.g. growing other varieties of gentian like spotted gentians (*Gentiana punctata*) or alpine plants like the hairy alpine rose (*Rhododendron hirsutum*), arnica (*Arnica montana*), alpine bellflowers (*Campanula alpina*), mountain pasque flowers (*Pulsatilla montana*), cowberries (*Vaccinium vitis-idaea*) or bilberries (*Vaccinium myrtillus*). There are symbiotic fungi in the soil, which the plants require to grow. If I were to sow these alpine plants at lower altitudes than they naturally grow, then the fungi would not be present. This is why I have to introduce native soil. When sowing these plants in their natural environments (at high altitudes and in alpine pastures) – e.g. with the intention of improving the population – this is of course no longer necessary.

Gentian seeds also need exposure to light to germinate. The seeds must not be covered or pushed into the soil, or they will rot. In nature the seeds just fall to the ground around the parent plant and are left exposed to the elements. Areas of churned up soil are good for germination. Soil is naturally churned up by cattle, sheep and deer. Once the seeds have been sown on the poorest and

In the foreground: yellow gentians (*Gentiana lutea*) in bloom.

Terrace with young gentian plants.

most barren soil possible, they will be left to their own resources completely – according to my experience it is not worth watering them or using fertiliser.

It is also important to pay attention to the difference in height between the location of the parent plant and the place the seeds will be sown: a 1,000m height difference will mean that the vegetation is so different that the plants 1,000m lower should be sown three to four weeks later. This is to compensate for the fact that the frosts come later at lower altitudes.

Many other plants can be propagated successfully using this technique. A large number of people who have attended my seminars and visited the Krameterhof have used my method successfully. After four to five years the gentian roots will reach a sufficient size for harvesting (over half a kilo fresh weight). Then they can be dug up and either used to make spirits (as a main ingredient or as a secondary ingredient) or dried for health or pharmaceutical purposes. Naturally I also harvest the plant seeds: as soon as the first seed pods begin to burst open, I cut the plants down and place them in paper bags. When the plants are dry the bags just need to be hit. The seeds fall to the bottom of the bag and the stems can then be pulled out.

Many animal and plant species are now endangered and gentians are one of them. The usual practice of nature conservationists is to put these animals and plants under 'protection', but they fail to take measures to preserve the habitat before it is too late. I would like to describe my own experiences in this area: forty years ago we had a large number of gentian roots in the mountain pasture. We even dug the old gentian roots up to a depth of half a metre with a 'root digger' – a 30-cm-long pointed shovel. When doing this the small secondary roots usually remained undamaged. The hole was filled back in with earth and stones so that the following year gentian seeds would be able to fall onto the churned, loosened soil and a thick growth of young plants could grow using the remaining secondary roots. The gentians grew very rapidly. Using this method they were rejuvenated.

For many years now it has been illegal for us not only to dig up gentian roots, but also to remove the plants or parts of them from our own land. It should be perfectly comprehensible that if the gentian roots are not dug up they get worn out from age (around 30 to 40 years), so the soil mats and compacts and the roots die out. However, I have worked with these plants, so I have the experience to know where and under which conditions gentians will grow healthily and propagate, how to get the seeds to germinate and I also know that they do well at lower altitudes with the help of symbiotic fungi. In my opinion it would make more sense if the 'protectors' of alpine plants (park rangers and, in Austria, the mountain rescue service) were trained to cultivate and propagate them, so that farmers would not have to be fined, just because they want to profit from these health-giving and valuable roots on their own mountain pastures. This would be a much more effective way to protect alpine flora.

Alternative Ways to Keep Livestock

Livestock play a large role in a permaculture system, they do not just provide high-quality produce, they are also industrious and pleasant workers. Poultry, pigs, horses, cattle, sheep, goats and many other animals can be used in a permaculture system. However, I only breed robust, hardy animals that are suited to the terrain. Old, sometimes rare domestic breeds as well as wild animals fulfil these requirements the best. I have successfully bred mouflon, chamois, ibexes and red, roe and fallow deer as well as various species of wild cattle such as yaks, water buffalo and American and European bison.

Most domestic animal breeds have become very rare. I consider it our duty to propagate them. In a time when the focus is always on high performance and this unhealthy drive for profit has affected the breeding and keeping of livestock, it is especially important to preserve diversity, which represents a significant cultural heritage. Old domestic breeds are not only significantly hardier, they are also more intelligent and adaptable than their over-bred and degenerate cousins. Their natural instincts are good enough for them not to fall over their own feet and their produce is of a far higher quality. These old breeds can be kept under near-natural conditions, which for me is the main requirement for keeping livestock.

Intensive livestock farming, on the other hand, amounts to torture of large numbers of animals. People should really ask themselves if they would want to be treated in such a way. In my opinion, the pain of animals is also felt by people. Anyone who has seen the inside of a battery farm will know what I mean. The fact that food produced in this way is not real food, but of very low nutritional value must be well known to a reasonable number of people by now. The use of growth hormones, antibiotics and sedatives in modern 'meat production' is very common. Since the BSE crisis the whole world has found out that there is nothing people will refrain from feeding to livestock. Also the enormous amounts of stress the poor creatures are subjected to when they are transported for days at a time is passed on as stress hormones in the meat. We are the ones that are being damaged most by all of these crimes committed against our fellow creatures. Fortunately, more and more people are beginning to question their own consumer behaviour. If there were no longer a market for these cheap products, the cruel treatment of animals in 'meat factories' would soon come to an end. Keeping livestock humanely, on the other hand, provides us with high-quality and delicious food, gives the animals as natural a life as possible and brings farmers joy every day!

On the Krameterhof, livestock are always kept outside in families and great care is taken to meet the needs of different animal species. As the livestock are always kept in family groups, the proportion of male and female animals also has to be adjusted appropriately. The size of the area available to them depends on the space naturally required by each species. With a little understanding and insight it is quite straightforward to have happy livestock on the farm.

Keeping livestock outside gives them a happy life.

Pigs in a Permaculture System

Pigs are an inseparable part of Holzer permaculture. By allowing my livestock to be free-range and using a system of paddocks, I reduce the work and the amount of feed required to a minimum. I also receive high quality food in the form of meat and bacon and I can sell on the young pigs that I breed. Finally, the animals actually work for me by loosening the soil and tilling my terraces. For me there is no more versatile or more helpful an animal, I only need to give them the right opportunity. If I were to close them away in a pen, however, I would have to work for my pigs instead. When keeping livestock, it should be the first priority of every farmer to give his animals a good life. In the end, they are the ones that provide the produce. Every farmer should be able to say when looking at their livestock that if they were to swap places they would also be happy.

Like all domesticated animals, pigs have lost many of their old character-istics over years of selective breeding. These 'high performance' breeds are no longer suited to being kept under natural conditions. They would trip over their own feet on rough terrain and they would scarcely survive the winter. In addition, they no longer have the natural instincts that they need to be good workers. For this reason I only keep old pig breeds on my farm. They fulfil all of the necessary criteria and are much more valuable, because they are only bred on a small scale.

Characteristics of a Few Old Pig Breeds

• Mangalitza

Mangalitza wooly pigs are medium- to large-framed with strong bones, powerful muscles and large lop ears. Their thick curly coat ('wool') is blonde, red or black-brown (swallow-bellied) and protects the pigs from the cold and rain very well. The brown and white stripes of the piglets makes them look similar to wild pigs. Mangalitzas were bred all over Europe for their excellent bacon up until the middle of the 19th century. They were gradually superseded as the production of meat became more intensive and today they are an endangered breed. They are very undemanding and are well suited to being free-range as long as their paddocks are large enough to satisfy their need to move around. The thick undercoat that protects the pigs so well against the cold is moulted in spring. If they have the opportunity to wallow, they can cope incredibly well with high temperatures. The maternal instincts of the females are particularly good. The animals are well known as bacon pigs with a thick layer of fat and high-quality meat.

Mangalitza pigs: their thick 'wool' protects them against the cold in winter very well.

A Swabian-Hall piglet takes a break from 'work'.

• Swabian-Hall Swine

Swabian-Hall swine are large-framed animals with a long body and lop ears. Their black and white colouring makes them difficult to confuse with other pig breeds. The animals are distinguished by their incredible good-naturedness. They are also extremely hardy and their meat is of a very high quality. Despite their excellent breeding qualities they are now endangered.

• Duroc

The duroc pig breed emerged in the north-eastern United States in the middle of the 19th century from crossing Spanish pig breeds. The animals have a reddish coat, are medium- to large-framed with an arched back and small lop ears. They are distinguished by their calmness and docility and their strong resistance to stress. Duroc pigs are also very hardy and sure-footed, which makes them good on difficult terrain. The fact that the quality of their meat is very high is also widely known.

Duroc with piglets in a roundwood shelter.

• Turopolje

The turopolje pig breed comes from what are, today, the pastures of the river Sava in Croatia. In the riverside woods of the Sava the animals were kept outside all year round. In order to adapt to this environment they became excellent swimmers. This allowed them to successfully find food in the large flooded meadows. Turopolje pigs are black and white spotted, medium-framed animals with large half-drooping ears. They are well suited to pastures and a high proportion of their diet comes from foraging. Their meat is also of a very high quality, however, like manga-litzas, they have a relatively large layer of fat. The turopolje is a critically endangered pig breed.

Free-roaming pigs are still a rarity in Austria and raise a great deal of public interest. The strikingly

Turopolje pigs are also a very hardy breed.

coloured and less well-known old breeds with their piglets especially fascinate visitors and passers-by. This positive impression encourages acceptance of and interest in endangered domestic breeds and makes it easier to directly market the products made from them.

Pigs as Helpers

Pigs make pleasant and helpful workers in many respects. The soil can be greatly improved by the pigs' digging activity. When they search for food they plough through the top layers of earth and loosen and aerate the soil.

It is easy to precisely direct the pigs by scattering loose feed (e.g. peas, grain or maize) in the appropriate places. Compacted soil can be loosened with minimal effort and well prepared for sowing afterwards. With this method I can direct the animals for small- to large-scale tilling. The pigs perform physically

'Staff meeting'

demanding work with ease. On the Krameterhof, where there is a great deal of rough terrain and the soil is so stony in some places that it cannot be ploughed, the pigs are indispensable as living ploughs.

When building paddocks I take note of how the pigs can help me best. Orchards and seedling nurseries make ideal places for them to work. In contrast to goats and sheep, they do not damage the fruit trees. There are often large amounts of windfall fruit in orchards, which can lead to the spread of fungus and mould. If pigs are put out in these areas to pasture at the right time, this danger can be averted to a great extent. As previously mentioned, properly directed pigs can be a great help when regulating rapidly-growing wild plants.

An agricultural research project in Germany has investigated the effect of large numbers of pigs (duroc and mangalitza breeds among others) on the plant kingdom when they were allowed to roam freely. The vegetation charting showed that the plant diversity in the areas grazed by the pigs doubled. The reason suggested is that plants growing in very overgrown areas that would otherwise not have the opportunity to develop suddenly had the chance to germinate and grow when the turf was broken.

Pigs make a large contribution to the regulation of snails. Movable pens can be used to place the pigs exactly where they are needed. Between the fields, which have a large population of snails, a narrow strip as long as the cultivated area can be enclosed. A fence of wire mesh is enough for short term use. Mobile shelters like horse trailers, transporters or anything similar make suitable open housing. The pigs must first get used to snails as a food source. To do this I mix

'Pig ploughs' of all ages 'working'.

snails I have collected in with their normal feed. The animals soon acquire a taste for them, start looking for snails themselves and eat all newcomers to the paddock straight away. Pigs need a great deal of water to digest snails. This must never be forgotten when regulating the snail population! Insects, which go through the early part of their life-cycle in the soil, eg cockchafers, chafers etc, are controlled by the pigs' rooting activity.

Pigs also serve as a good example of the cyclical nature of permaculture systems: the soil is prepared and fertilised by the pigs, plants grow lush and healthily in it, windfall fruit and roots remaining in the soil serve as feed – at the same time snails and unwanted insects like cockchafer larvae are eaten – and last of all I have the finest bacon from humanely kept animals.

Keeping Pigs Using a Paddock System

On the Krameterhof, all of the land is managed as part of a paddock system. This means that all of the livestock are kept outside the whole year round. Animals that have their natural needs fulfilled remain healthy and happy, grow well and provide good offspring. Having enough space to move and dig around in is as important for the animals' welfare as giving them the chance to wallow and build nests so that they can bring their young safely into the world. Pigs do not have sweat glands, so they have to rely on bodies of water and wallowing to regulate their body temperature in hot weather. When they wallow, the pigs cover themselves in a layer of mud, which protects the light-skinned breeds

Muddy pools are particularly important for pigs.

and those with less hair from sunburn and makes a significant contribution to preventing parasitic infestation. Wet areas and flowing water are best for the pigs to make into their own muddy pools.

The amount of work for the farmer is minimised by keeping pigs in this way. When keeping pigs outdoors it is vital to take soil conditions and any hilly areas into account. The land must not become overused. It is important to ensure a correct stocking rate and that the pigs do not graze for too long. Continually observing the development of the pig population and the pasture areas will make it easy to prevent any damage from being caused. The stocking rate of the pigs should be adapted to fit the amount of food naturally available in the paddocks. Depending on soil conditions and vegetation, I keep between three and twelve pigs per hectare in a paddock. Simple open structures made of rough timber logs or stones are built for shelter. When choosing the right place to put one of these shelters, you need to observe and understand the pigs. Soon after they have been moved to a new paddock their favourite places to lie can be clearly ascertained. From what I have been able to observe, pigs are very sensitive to earth energies. In the places where the animals particularly like to rest I build one or more shelters – depending on the number of animals and their requirements.

Extra feeding is rarely necessary, because in a paddock system there is enough vegetation throughout the year and the pigs are kept outside all year round. Even in winter they find enough food beneath the snow – the pigs like to dig up Jerusalem artichokes, which taste like sweet potato.

When the pigs move from one paddock to the next, the churned up soil is sown with a mixture of different crops (turnips, potatoes, cabbages, peas etc.). In the next paddock I then use the pigs to do other things, such as reducing the number of stinging nettles between the fruit trees. This paddock will also be sown with a mixture of seeds after it has been grazed. Then there are further paddocks and the cycle continues. Once enough time has passed, the

Mixed herd of piglets in front of a roundwood shelter.

pigs can return to the first paddock. The system moves in a circle, which makes additional feeding unnecessary, because the animals work for themselves. I always try to make sure that enough tubers and root crops remain in the soil despite the grazing, so that the plants can propagate themselves. With Jerusalem artichokes the pigs' digging behaviour is particularly helpful, because the soil is not only loosened; it also spreads the tubers. Once the area has been grazed, the conditions are even better for propagation, regeneration and growth than in 'untilled' soil. Naturally, the paddocks are simultaneously working as arable land in a permaculture system, because after the pigs have done their work, the areas are used to cultivate crops. Therefore, paddocks are not a waste of land, but are in fact the most productive way of using it.

Wild Cattle and Old Domestic Cattle Breeds

For many years I have successfully bred old domestic and wild cattle breeds on the Krameterhof. Over the course of the years the composition and number of the most common breeds here have changed. I did most of my cattle breeding at the beginning of the 90s, when I kept a mixed herd of around 50 wild cattle in a 25-hectare paddock. The cattle I have bred over the years are European bison (*Bison bonasus*), American bison (*Bison bison*), yaks (*Bos Poephagus mutus*), water buffalo (*Bubalus Bubalus arnee*), and also domestic breeds such as Scottish Highland cattle, Hungarian steppe cattle and Dahomey miniature cattle. They continue to be a part of my breeding programme today.

American bison and Highland cow in a paddock together.

A yak's thick hair makes it very well suited to cold conditions.

For me keeping wild cattle is, above all, about propagating and maintaining the breed. To this end I am working in cooperation with a couple of zoos to breed European bison, also called wisent, which are critically endangered. Wild cattle also provide excellent high quality food. For example, yak and water buffalo milk and meat are a rare delicacy. All of these animals are particularly hardy and undemanding, which makes the work required to keep them very low. Happy livestock, high-quality produce and not focusing on high yields are both the results of and the key to success.

Keeping Bovine Species

I always keep my cattle using a system of paddocks. With this method the land is never overused and the soil and vegetation always have enough time to recover. A paddock system is particularly important when keeping cattle, because their weight can cause the soil to become permanently compacted. On the Krameterhof, bovine species form small herds and animals of different species can develop a bond. Only the water buffalo isolate themselves from the group and tend to stay close to the water. They prefer one particular pond. I think that it is very important to pay particular attention to the social relationships between the animals. If you get the composition of the group wrong it can, as with all livestock, lead to altercations. The animals develop a natural hierarchy, which is why you should make sure that there is a dominant bull in the group. The bull's rivals should all be significantly weaker. Organising a herd like this will help to prevent any serious fighting between rivals. I also arrange it so that the enclosures are large enough to provide them with plenty of space to avoid each other. Creating places to rest and hide is vital. Visual barriers in the form of wooded areas and hedges play a large part in this. The animals should not be put on display so that they can be seen from all sides as they are in many zoos. It is very important for the animals to have plenty of places to escape to and for them to be disturbed by people as little as possible. Wild animals can be kept in an enclosure without any problems as long as they are kept as naturally as possible and retain their wild animal character. In order for this to happen you must, of course, familiarise yourself with the way they live. In contrast to their overbred cousins, wild cattle know their own abilities very well, so the fencing around the paddock must be as well thought out as possible. Whether the animals are generally happy or not, a normal fence will not keep them in. Curiosity and their instinct to play alone will allow them to overcome these barriers very quickly. I have had the best results with two-metre-high electrified game fences.

Feed

I limit additional feeding to during the winter, because I use a paddock system. Like the pigs, the wild cattle also have lush forage fields at their disposal, in which they can find their winter food in the form of turnips, fodder kale, Jerusalem

Dahomey
miniature cattle
and Highland cow
with calf.

artichokes and many other plants. They also get hay, grain and apple and pear pomace, which is left over from making juice and cider on the farm.

I have always found it interesting to observe which plants the animals prefer when they change paddocks. Through lengthy observation I have determined that animals with diarrhoea caused by parasites in their stomachs or intestines eat poisonous plants such as lupins, monkshood, male fern, buttercups and even poisonous mushrooms.

By observing throughout the year I have been able to establish that animals which have a lush flora with a variety of poisonous mushrooms available to them stay healthy. They also no longer need to be wormed. The district vet from Murau, Dr. Fritz Rossian, has been examining our livestock and issuing the necessary health certificates for the Tamsweg district council for decades. Dr Rossian was very enthusiastic about this system. I have had the same success with other livestock. You only have to make sure that these plants are present in large enough numbers and are richly diverse. The animals can use their instincts to decide for themselves when and which plants they require. Obviously, they should not be put in a position where they have to eat these plants out of hunger, because they cannot find anything else. For this reason you should under no circumstances mix herbs or poisonous plants into the feed yourself. Only the animals know what they need. Unfortunately, I have never found the time to work out the dosages that would be required. This would be an important area for scientific research. That which is applicable to animals is also applicable to humans. Food is also medicine! It must be varied, nutritious and healthy, which means it must be free from artificial additives and not contaminated by fertilisers or pesticides.

Finally, I would also like to say something about my method of keeping cattle in general: to be more specific, dehorning. It is unbelievably painful for the animals and also has an effect on their behaviour. According to my observations they act in a completely different and disturbed way. They butt each other in the stomach, which can lead to premature or stillbirths in pregnant cows. In addition to this, I am of the opinion that dehorning cattle also affects them in other ways. I think it is possible that animals also store and dispose of harmful substances in their claws and horns. Dehorning as well as docking tails and cropping ears is nothing more than mutilation as far as I am concerned. I am convinced that we should be held to account for the way we treat animals. If we keep this in mind and treat our fellow creatures with consideration, keeping livestock will bring us joy and success.

Poultry

Bird Conservation

Before I move on to keeping and breeding poultry, I would like to point out the value and importance of local bird species. Numerous bird species are now endangered as a result of the loss of their natural habitats. The draining of meadows and wetland, river regulation, land levelling, unhealthy agricultural methods and the increasing use of pesticides are causing the red list of threatened and endangered species to grow each year. Urban sprawl and an ever denser traffic network continue to contribute to the loss of natural habitats. I see it as my duty to help to change this sad situation with my own farming practices. Birds play a valuable role in controlling the insect population and help to propagate and seed numerous plants. Birds are incredibly useful and beneficial creatures and should be supported by every means available.

Even those with small gardens can make a valuable contribution to bird conservation. Well-structured gardens as opposed to neat lawns and hedges of fruit bushes instead of monocultures offer birds a habitat and a source of food. The use of chemicals must be abandoned, so that the birds' natural source of food is not poisoned! The greater the diversity of plant varieties within the hedges, in the meadows or forests, the greater the diversity of the yield of fruits and berries. Then a greater variety of insects will also start to appear. This will ensure a balanced diet. Insectivorous birds like robins (*Erithacus rubecula*) and wrens (*Troglodytes troglodytes*) will find a generous buffet and the population of beetles, butterflies, greenflies and whiteflies will never become large enough to cause any damage. Good forage plants for local bird species are the elderberry (*Sambucus nigra*), Guelder rose (*Viburnum opulus*), wayfaring tree (*Viburnum lantana*), wild cherry (*Prunus avium*), bird cherry (*Prunus padus*), fly honeysuckle (*Lonicera xylosteum*), barberry (*Berberis vulgaris*), bramble (*Rubus fruticosus*), dog rose (*Rosa canina*), wild privet (*Ligustrum vulgare*), yew (*Taxus baccata*), ivy (*Hedera helix*), spindle tree (*Euonymus europaeus*), dogwood (*Cornus sanguinea*), snowy mespilus (*Amelanchier ovalis*), whitebeam (*Sorbus aria*), rowan (*Sorbus aucuparia*) and many others. These trees and shrubs provide the birds with a varied diet of berries, fruit and seeds. Once an adequate range of food has been ensured, it is time to create nesting sites. Open nesting birds particularly like to nest in dense thorny hedges. Holzer permaculture provides cavity nesters with many hollow old trees. Next, a number of different nest boxes are prepared. The sizes of the boxes and of the entrance holes should be varied, so that not only a small number of more competitive species like the great tit (*Parus major*) are encouraged.

The birds do not need to be fed over the winter, because they will find enough food amongst the diverse plant life of a permaculture system even in winter. However, I have still planted a few forage plants for the birds close to the house. This makes it possible to watch the birds closer up and learn more about their behaviour and food preferences. The forage plants must be properly selected and maintained. They should be planted in sheltered places, such as next to trees. The food should contain seeds of different sizes (sunflower seeds, flax seeds, millet and hemp seeds), so that more than just a few types of bird are attracted. It is particularly important to make sure that the feed does not become wet and is not contaminated by droppings. Unsuitable winter feed can lead to the spread of diseases and parasites! In order to observe woodpeckers (*Picidae*), treecreepers (*Certhiidae*) and nuthatches (*Sitta europaea*) a mixture of fat and feed can be spread onto trees (in cracks in the bark) in winter. Another possibility for winter feeding is cutting off the dried seed heads of indigenous shrubs and bushes in autumn. The seed heads can be hung up in winter and will provide the birds with ample food. During winter, birds should only be fed when the ground is completely covered with snow. However, the best thing that you can do for the birds in your garden, is to provide them with good food by not harvesting everything in your permaculture system. Supposedly untidy things like piles of brushwood and thick hedges are well-valued by birds: in winter they can also find insects and other small creatures there!

Keeping Poultry Humanely

We keep birds extensively on the Krameterhof and we are mostly self-sufficient. It is, of course, possible to use poultry breeding as a source of income for a farm without resorting to inhumane methods.

When keeping poultry I try to understand the birds' natural habitat and to reduce the need for additional feeding by introducing selected forage plants. The birds should be able to live as independently as possible and provide high quality produce under the best conditions. Birds which have hatched naturally are the most suitable for breeding. Incubator hatched birds do not have the maternal instincts needed to be able to raise a brood independently. Often they do not brood for long enough, leave the nest too early or do not look after their young properly. Acquiring naturally hatched birds is not always easy, but it is worth it. Many rare bird enthusiasts and breeders still believe birds that are naturally hatched and kept free range should not be left to hatch their own eggs. They check regularly for eggs and if any are found they catch the hen, duck or goose and remove the eggs. The eggs are kept warm for a day and then returned. I am convinced that this behaviour hurts the birds' natural breeding and brooding instinct more than it helps it. Of course, sometimes a mother bird will not hatch her eggs even under optimal conditions. However, I accept this as genetic selection. The successful mothers balance out the losses. In the course of time, the desired number of reliable breeding pairs will establish itself. I have had particularly good results with mallards and Indian runner and crested ducks as well as Styrian chickens. These breeds are hardy and adaptable and help to keep the snail population down.

To protect my poultry from predators I plant hedges. For these protective hedges, which I might for instance put in a chicken enclosure, I use a number of varieties of very thorny plants. Thorn hedges made of different varieties of rose are particularly suitable. For one thing, they fulfil their function as a place for the birds to take shelter, they are popular with the birds and, furthermore, the fruits of the different roses provide them with a tasty food source. In addition to that, I enjoy the beautiful flowers and the heady perfume of the rose hedges every day. The following wild roses are recommended: the multiflora rose (*Rosa multiflora*), a strong climber, has abundant flowers and an intense perfume, it is also very popular with bees; the dog rose (*Rosa canina*), an easy to cultivate wild rose and also an excellent medicinal plant, its fruit is rich in vitamin C and also makes delicious jam or fruit tea; and the Japanese rose (*Rosa rugosa*) is also very suitable for a rose hedge of this kind. In Austria it is called the 'apple rose', which probably comes from the plant's large red shiny fruit (rose hips), which are very tasty. It is also known as the 'potato rose', which comes from its corrugated leaves.

I construct mobile nesting sites for the poultry. They consist of two pieces of rough timber positioned in the thorny undergrowth in such a way that there is still enough space for a hen and her clutch between the pieces of wood. The

Muscovy ducks on the Krameterhof.

A capercaillie once even performed a courtship display on my arm.

advantage of this kind of nesting site is that they can be put together in different enclosures if needed. In my experience, mobile nest sites are hardly ever attacked by predators. Moving the sites seems to repel the distrustful predators. The thorns present an additional barrier.

Over the years, building ponds for breeding ducks and geese has proven to be particularly worthwhile. For this purpose, I have created islands of differing sizes which can only be reached from the bank using rickety wooden constructions (usually a single plank is enough). Predators like foxes and martens shy away from the water and shrink back from the unstable entrance to the island over a bridge. On the island there are sheltered nesting sites and the area is planted with different varieties of willow. This will also provide protection from birds of prey. There should be twice as many nesting sites as there are ducks capable of brooding, to ensure good breeding results. The nesting sites should be dry and well ventilated (but not draughty). In my experience, the birds prefer partially dark, secluded places to nest. It is especially important at the beginning of a new breeding season that these sites can be closed off. The birds need to get used to their new habitat slowly. The ducks and geese must have access to stretches of open water all year round. If the pond freezes over completely in winter, they will be left defenceless against predators. For this reason the inflow to the pond should be at as steep an angle as possible. The water pressure will then keep the area of water around the inflow free of ice.

Ducks are omnivores, their diet ranges from young leaves, roots, aquatic plants and grain to worms, amphibians and even small fish. They also really like to eat snails! Geese live exclusively on plants. They particularly like to eat grass and meadow plants. They graze and fertilise small areas of meadow and are also the best alarm system on the farm. As a result of their strong territorial behaviour they announce visitors with their loud honking. As they are unerring in this respect, they have long been used as guard animals worldwide.

Over the years I have also had experience breeding quail, pheasants and other wild poultry. If the needs of these wild birds are successfully met it is even possible to breed very demanding native grouse like the capercaillie (*Tetrao urogallus*) and hazel grouse (*Bonasa bonasia*). With enough sympathy and understanding almost anything is possible.

Earth Cellars and Open Shelters

As the animals on the Krameterhof live outside all year round, building simple and reliable shelters is especially important. The animals need dry places to retreat to that are sheltered from the wind. I try to use as economical a building method as I can. As I have said, the winters in Lungau are very harsh with temperatures sometimes dropping below -25ºC, so I decided to put the shelters underground. I make use of the soil's ability to insulate and store heat. By doing this, I can build the shelters much more simply and cheaply and, because they are earth sheltered, create a draught-free and warm shelter, in which all livestock can be comfortable. These shelters can be built differently to suit different types of animal and the length of time they will be kept there. Over the years, I have developed some very simple and efficient methods of building shelters, which I will now explain in greater detail.

Earth Shelters as Pigsties

Building simple earth shelters to house pigs is very straight forward. These buildings are put together with the smallest effort and fulfil all of the important requirements for pigs. First of all, a two- to three-metre-wide and one- to two-metre-deep ditch is dug at the foot of a slope. The location must, of course, be

Earth shelter with mangalitza pigs.

selected so that the ditch remains dry. Next, tree trunks are laid over the ditch to form a roof. The roof should be gently sloped so that any water will drain straight off it. To insulate the roof, sheeting or bitumen shingles are placed on top of it and the whole thing is covered with a little soil or brushwood. The length as well as the width of the building can be increased, but the logs for the roof will have to be thicker. The width, however, should not be more than four metres for statistical reasons. As the trunks are only lying on the soil and the pigs tend to dig a little around them, they must have an excess length of at least one metre. The ditch must have low side walls for the same reason. This will ensure stability. The building only has one side open, so that the pigs can go in and out. Now all you need to do is throw straw or hay into the entrance and the pigs' luxury accommodation is complete.

These earth shelters do not have to be 'easily accessible' as far as I am concerned, because the pigs do not need me. I only need to throw the straw into the entrance and the animals will distribute it themselves. Cleaning out dung is also unnecessary, because pigs are very clean animals. They go outside to relieve themselves and keep their 'home' clean. The reputation pigs have for being 'dirty' only comes from the fact that they cannot move about in cramped pens, which forces them to live in their own muck. If they have the option to move, however, their sleeping places are wonderfully clean. Many people marvel at this fact again and again when they are shown around. They say that they would have no reservations about sleeping in a pigsty after that.

Roundwood Shelters and Earth Cellars

With larger building projects, I think it is particularly important that I keep as many alternative uses available for my buildings and structures as possible. I try to build them in a way that can be used for very different purposes with very little or no alterations. This way I prevent the need for large-scale building alterations

Roundwood shelter made of spruce and larch.

and the annoyance that comes with them right from the start. Now I also build most roundwood shelters larger and I make them accessible to vehicles. This is so I can keep other livestock like cattle or horses in them with no difficulty. I can also use these buildings as storage rooms with only minor alterations.

The shelter is designed so that it is closed on three sides and that the entrance faces east. This way the shelter faces the rising sun and the first rays of sunlight will wake the animals. If it is too hot for the animals at midday during the summer, they are sheltered from the sun inside and have a pleasant cool living temperature.

To build the shelter I begin by digging into a terrace. That way I do not have to dig down, but just dig into the slope. Using this technique saves a great deal of digging. Once the area for the shelter has been dug out, the excavator digs a narrow ditch at the back and to the sides approximately one metre deep. The logs for the walls will be placed in this ditch. During this stage in the construction the ditch remains open. To make each wall even, it is necessary to place all the logs and fill around them in one go. The logs can be leant against the inside of the slope in the meantime. This work is easily carried out with an excavator. The majority of excavators have grappler arms as additional attachments, which are very well suited to this work. If one is not available, the logs can simply be attached to the excavator with a belt or chain and lifted into the ditch. When choosing timber you should make sure that the dimensions are large enough, because the thicker the logs, the longer they will last. The type of wood also plays a large role in this. Larch and robinia last the longest. The quality of the timber, in other words factors such as how knotty it is or whether it is infested with bark beetles, is, however, of minor importance. This allows you to make good use of low-quality and therefore low-priced timber. I have constructed many of my roundwood shelters out of timber from windbreak as I described in the chapter 'Landscape Design'. Once a wall has been put up the ditch is filled in and the excavator is used to arrange the logs. With a two-metre-high roundwood shelter the logs have to reach one metre into the soil. So you must use timber that is at least three metres long. When all three walls are completed, the logs are all cut to the same height. This is necessary, because it is not always possible to put in all of the trunks evenly, so you should allow a little leeway in the length of the timber. The back wall should be somewhat higher, so that it can fit closely with the top of the roof. This will increase stability even more. After a wall has been cut to the same height, the individual logs are held together with a cap log. To do this, a log is cut in two lengthways, laid on top of the wall and nailed on. Once this is done for both the side walls, work on the roof can begin. Logs are lifted onto the shelter with an excavator. These are then notched with a chainsaw and nailed to the walls. The notches improve the fit between the roof and walls and prevent the shelter from giving way at the sides as well as improving general stability.

Finally, the roof is covered with sheeting or roofing membrane and earth is heaped over it. The best way to do this is the following: first of all, I place the membrane on the logs. Next, a little earth can be put on the membrane. This is to fill in the spaces between the logs a little. Making the roof more even also helps to direct surface run-off. Now the sheeting goes on top. At this point it is important to make sure the sheeting is not damaged, so the earth underneath the sheeting must under no circumstances contain any stones. Pond liner makes the most suitable sheeting, because it is by far the most hard-wearing. Finally, an additional layer of membrane is placed on top of the sheeting for security. Neither

Scottish Highland cow and Dahomey miniature cow in a roundwood shelter.

the roofing membrane nor the sheeting should be taut, otherwise they could tear under the weight of the very heavy load. Now the shelter is covered. The depth it is covered to will vary depending on the soil. It should not measure more than half a metre. Loam soil is particularly heavy, so less should be used. Whilst covering the shelter with earth, soil should, of course, be packed around the side walls as well. The best way to round out the whole building is to put buttresses against the sides. They provide the soil on each side with additional stability.

The size and design of these shelters can of course vary, but the width should not go beyond four metres, because of the load on the roof. However, there are few limits on the depth. I use rough timber logs, because they reduce the work load, guarantee greater stability and ensure that the structure will last longer. Naturally, the construction can also be made from finished timber, but I consider my method to be much easier and cheaper. It is possible to build a roundwood shelter of this kind with a width of three metres, a length of six metres and a height of two metres with an excavator and one additional person within a day. That makes this one of the fastest and cheapest methods of building an insulated shelter or earth cellar. The cost of renting the excavator comes to around 400 to 500 euros a day in Austria (around 10 work hours). If you allow for the cost of the sheeting, nails and your own labour, the total will be between 700 and 800 euros, if you have the timber already. The lifespan of a building like this, of course, varies depending on the type of wood used and the dimensions. If you use larch or robinia wood with a diameter of 30 to 40cm, it will last for around 30 years. For a building constructed in a single day that is a remarkably long time.

A shelter of this kind is very well suited to cattle because of its height. As cattle are not as clean as pigs, the shelter must be easily accessible. I have solved this problem by building it three metres high so that I can easily muck it out with a tractor. The costs are low, less work is required and the livestock are happy. That is the perfect combination for me.

Use as a Storage Room

Earth is the cheapest and best insulation. The temperature of the soil balances out fluctuations and provides a steady room temperature, which is not just good for livestock, it means the shelters can also be used to store fruit and crops. As my paddocks are also used to grow crops at the same time and the animals change paddocks regularly, some of the shelters can be used as storage rooms during the winter – and they are exactly where the crops are. So if I want to store turnips or potatoes, for example, I only have to remove the straw or hay with the tractor fork and, if necessary, pour in a tractor bucket of sand. After this small amount of work I just need to tip the crops in with the tractor. To insulate the building for storage through winter I block the entrance with straw. The insulation coupled with the temperature of the soil will stop the building from freezing in even the harshest of winters. If I only need the building for storage, it is easy to build a door and create an easily accessible storage room.

BUILDING A ROUNDWOOD SHELTER OR EARTH CELLAR

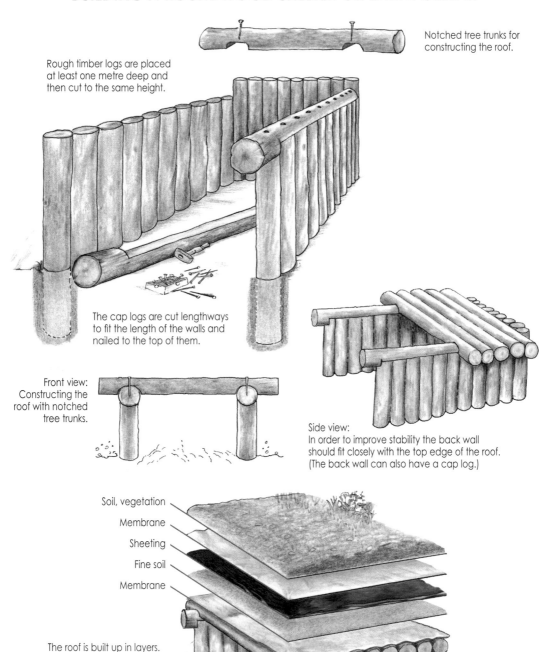

Notched tree trunks for constructing the roof.

Rough timber logs are placed at least one metre deep and then cut to the same height.

The cap logs are cut lengthways to fit the length of the walls and nailed to the top of them.

Front view: Constructing the roof with notched tree trunks.

Side view:
In order to improve stability the back wall should fit closely with the top edge of the roof. (The back wall can also have a cap log.)

Soil, vegetation
Membrane
Sheeting
Fine soil
Membrane

The roof is built up in layers.

Storage room made from robinia logs.

In addition to balancing out the temperature, the high levels of humidity are an important advantage of these storage rooms. These days many houses have cellars with heating systems built into them. Concrete floors also replaced the compacted earth floors in cellars a long time ago. However, this is disastrous for storing fruit and vegetables. The heating means that the humidity is usually so low that apples put into storage wither and wrinkle in the shortest amount of time. However, the high humidity, around 80 to 90 percent in earth cellars, and their fairly steady temperature between 8 and 10ºC is ideal for storing the majority of crops.

Stone Cellars

I have also constructed a few stone cellars specially for storing fruit and vegeables on the Krameterhof. They have the same basic characteristics as a cellar made from wood, only stone cellars are meant to last for ever. This means that building one requires a great deal more work.

When building a storage room just for fruit and vegetables, it is a good idea to take a number of details into account. Good air circulation is important. Gravel is laid on the floor for drainage. The ventilation pipes must be large enough to provide the room with the required amount of oxygen. The air goes through a ten-metre-long underground pipe into my stone cellar. The pipe reaches to a depth of around one metre. On its way through the pipe the air is brought to the same temperature as the soil. If the temperature of the air coming in was

Stone cellar on the Krameterhof.

Stone cellar for storing fruit.

STONE CELLAR

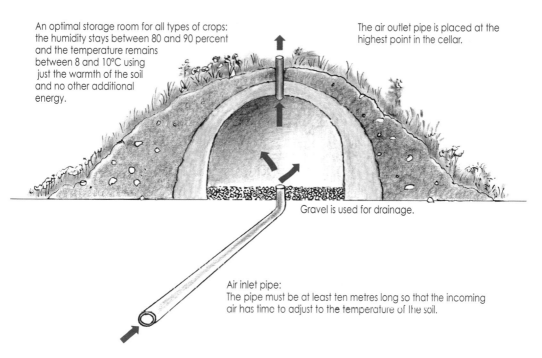

An optimal storage room for all types of crops: the humidity stays between 80 and 90 percent and the temperature remains between 8 and 10°C using just the warmth of the soil and no other additional energy.

The air outlet pipe is placed at the highest point in the cellar.

Gravel is used for drainage.

Air inlet pipe:
The pipe must be at least ten metres long so that the incoming air has time to adjust to the temperature of the soil.

not the same as the air inside, there could be an unwanted temperature drop in the cellar. The result would be a build-up of condensation. The dripping water could make the crops begin to rot or moulder. Even the diameter of the air inlet and outlet pipes must be adjusted according to the size of the room. For a 100m³ earth cellar, for example, I use ventilation pipes with a diameter of at least 15cm.

If the air inlet pipe is sloped slightly downwards, it can also be used as a drain pipe when the cellar is being cleaned. The air outlet pipe should be put in at the highest point in the room. This also helps to prevent condensation from building up.

Naturally, roundwood shelters and storage rooms must be checked regularly like any other building. This is necessary in order to repair any damage promptly and to be able to guarantee safety.

Building roundwood shelters and earth cellars is both possible and practical all over the world. I have had good results with all of my projects. If you make use of the balancing effect of the soil, you can create a pleasant, cool retreat in hot places and a warm one in cold places. You just have to understand how to make use of nature's resources properly.

3 *Fruit Trees*

Possible Uses

Fruit trees and bushes fulfil a number of functions within my permaculture system. They provide vitamin-rich, healthy food, which can be processed to create many different products such as jams, preserves, juices, vinegar, wine, spirits etc. Fruit trees are also very well suited to growing in paddocks where livestock is kept, because they are an excellent source of food. The windfall fruit make an especially good source of high quality feed for pigs. The fruit blossoms provide a great number of insects with a rich source of food. Bees, which play a substantial role in pollinating fruit trees, particularly benefit from the fruit blossoms. If there are enough bee colonies nearby, the number of pollinated flowers will increase dramatically along with the size of the yield. Wood from fruit trees (mostly wild fruit trees), and especially from pear and cherry trees, is highly valued as top quality joinery and industrial timber. The roots are very popular with artists, because the burl wood can be used to make unique and beautifully shaped objects such as wood carvings or other works of art. The aesthetic value of the trees should also be considered. From spring to winter, a blossoming, fragrant orchard in fruit brings joy to the heart and soul every time you walk through it.

Planting fruit trees costs me no more than planting any other kind of tree. The costs are not higher when I plant an orchard, because I just sow the trees and then graft the varieties I want to grow. If you use this method, you will need to have a great deal of patience, because it will take a long time for the trees to give their first yield. As fruit trees have so many advantages, I try to grow as many of them on my land as possible. I can use all of my terraces for growing fruit trees, crops and for keeping livestock at the same time.

I also plant cultivated and wild fruit trees in the forest to increase the diversity of species there and to increase the range of functions available for my woodland plots. From my point of view, there is no reason not to simply plant fruit trees (cultivated and wild) together in a mixed culture in the forest. In the state of Tirol, at least, the authorities agree with me on this point: a young farmer told me about a business consultation in which the introduction of wild fruit trees was being actively encouraged near the town of Kufstein using the striking tag line 'jewels of the forest'. If all authorities were of the same opinion, I could have saved myself many disputes and a great deal of time and frustration.

On the Krameterhof there are several thousand fruit trees of different varieties and sizes. Selling fruit trees and bushes has been one of the farm's

Fruit trees in blossom are not just pleasing to the eye;
they also make excellent honey plants.

most important sources of income for some time. I have overseen the design and planting of many gardens, recreational areas and public grounds (from parks and playgrounds to cemeteries). As my trees are not sprayed with pesticides, fertilised, watered or pruned, they must develop into hardy and independent trees in order to grow and thrive under these conditions. Over the course of time, they have also adapted themselves well to the climatic conditions in Lungau where there are large temperature differences during the day and at night and a greater danger of frost. This is why I have been able to guarantee that my trees would grow well the year they were planted and the next without taking any great risk. I knew that failure would be unlikely as long as I planted the trees myself and the owner followed my instructions: leave the trees alone as much as possible and do not tend them excessively. This guarantee has given me a large commercial advantage, because I was the only one who was able to give a guarantee of this kind and the only one to agree to replace any trees that did not grow. These days I do not have the time to sell trees individually or to take on small planting jobs, despite the demand still being very high. Time constraints mean that I can only oversee a few of the larger and more interesting projects, which my plants are best suited for.

Many fruit trees have been planted throughout the Krameterhof. They have also been planted on steep and rocky terrain, because they help to stabilise the slope with their deep roots, which provides valuable security. Naturally, these trees are not for sale, instead they stay where they have been planted. The fruit

Terrace with pear trees that is also used to grow cereal crops (ancient rye).

Cherry trees next to rowan trees, spruces, larches and Swiss pines.

here is either harvested, if the land is easily accessible, we have the time or there is demand for it, or it falls from the tree and provides the animals with food. On these areas of land old and rare fruit varieties, which generate a lot of interest from distilleries, are generally grown. My Subira pears, for example, are much sought after in the production of schnapps. When I sell them I arrange that the buyer harvests the pears themselves. This means that I only have to give the distilleries the correct harvesting time. Then they send people to harvest the fruit and still pay a good price for this hard-to-find pear variety. At this altitude they develop a very intense flavour, which enhances the quality and taste of the schnapps.

A healthy mixture of cultivated and wild fruit trees grow throughout the Krameterhof. Wild fruit trees can pollinate many cultivated fruit trees. Wild fruit trees are very good for making schnapps and vinegar. They can also be used for making jams and juices and for medicinal purposes. I particularly like to plant:

- Crab apple (*Malus sylvestris*)
- Wild pear (*Pyrus pyraster*)
- Wild cherry (*Prunus avium*)
- Blackthorn (*Prunus spinosa*)
- Rowan (*Sorbus aucuparia*)
- Wild service (*Sorbus torminalis*)
- Service tree (*Sorbus domestica*)
- Cornelian cherry (*Cornus mas*)
- Snowy mespilus (*Amelanchier ovalis*)

View of fruit tree terraces on the Krameterhof. The different flowering times prevents a complete crop failure from late frosts.

107

As the trees grow in a mixed culture, they blossom at different times, which means that a complete crop failure is prevented if the climatic conditions are unfavourable (late frost). This diversity and the different flowering times ensures that there is plenty of pollen available for the blossoms to be pollinated, which will ensure a good yield of fruit.

The Wrong Way to Cultivate Fruit Trees

From my childhood to this day I have sown, planted and tended thousands of trees. Even as a child I was sorry for every twig I had to cut. This meant that I was always a little negligent when it came to pruning my trees. As a result, a sort of wilderness developed in my first garden, the *Beißwurmboanling*, over a period of time. During my training to become an arborist at agricultural school, I learnt that fruit trees should supposedly be pruned, fertilised and sprayed with pesticides in order for them to grow well. We were also shown how to catch, poison and gas voles to prevent them from damaging the fruit trees. It is shocking that these practices are still described in almost all textbooks today.

During my training I also learnt about the conventional view of how fruit trees should be planted: for instance, we dug a hole for a fruit tree measuring one metre across and 40-50 cm deep. Then we put in galvanised wire mesh, which was bent at the edges to keep the voles away from the roots, and filled in earth around the tree. The earth was first mixed with a shovel's worth of chemical fertiliser and a great deal of water. After that, we hammered in a stake and tied the tree to it with a leather strap in a figure of eight. Then branches at an angle of less than 45 degrees from the trunk for apple trees and a 60-degree angle for pear trees were removed. We cut each branch back to an outward-facing bud. This is intended to make the branches grow outwards from the tree. Most of the inner branches were also removed, so that more sunlight could reach the crown. This is meant to help the fruit to develop better. The branches of pear trees are at an angle of 60 degrees, because they have deep roots, which means that the tree can support branches at a steeper angle. The sturdy stake is meant to stabilise the tree so that it is kept upright and grows better. Wind and snow will not bring it down so easily. This sounds quite plausible. Poisoning and gassing voles also makes sense – they eat the tree roots. They will, however, do this in spite of the wire mesh, because they often dig straight down from above. It is also possible that in time the mesh will rust and no longer be capable of protecting the roots.

The intensive use of fertiliser when the tree is planted is supposed to help it grow. The pesticides to protect the tree from fungal diseases and many different kinds of 'pests' are, according to experts, also justified. The economical argument should be clear to any layperson. The measures described here to 'tend' fruit trees require a great deal of energy and ensure that the trees will continue to need constant care. Trees cultivated in this way are dependent on human care from the first day. They are 'addicted' to regular fertilising and watering and they are susceptible to scab, fungus and frost. They are easily damaged by the wind or snow and they are vulnerable to 'pests' of every kind. According to the rules of these conventional methods, cultivating fruit trees at high altitudes should not be possible at all.

When I finished my training, I received a certificate to allow me to purchase the strongest poisons like parathion. We used these poisons during training. Once I had internalised the conventional methods, I became almost ashamed that I had a wilderness garden at home. So I practised what I had been taught immediately and 'tidied it up'. The trees were pruned, sprayed with pesticides and fertilised. I bought chemical fertiliser and I bought mouse poison and gas from the chemist's by the kilogram. I removed the turf in a radius of one meter around each tree, hammered in stakes and tied the trees to them tightly. I cut the espalier trees on the wall of the house back vigorously and bound them tightly to the frame. I continued to remove any grass that grew within one meter of each tree. The voles that I could not reach with poison or gas, I gassed with my moped by directing the fumes from the exhaust pipe through a tube into the entrance of the voles' burrows. This was also a recommendation given by my school. I carried out these tasks for a whole year with great energy. I also used these methods on my customers' fruit trees, of which there were many. The following year, I discovered that all of my espalier trees were in a pitiful state. Although a few of them had put out new shoots at the sides or close to the ground, the apricot and peach trees were no longer showing any signs of life. I despaired, because I could not work out what the cause of all of this damage was. I had done everything according to the textbook! When I visited my customers in the spring to sell to them and to prune, fertilise and spray the fruit trees with pesticides, it suddenly occurred to me. At the Schuster-Bartl's farm in Ramingstein, with whom I have had

Healthy tree in my 'wilderness garden'.

good business for many years and who have always welcomed me, suddenly the reception was very cold. Mrs Schuster-Bartl – she was well known for being a very strong-willed farmer – greeted me with the words: "Aha, here he comes! You're the one who ruined everything with your chemical fertiliser and pruning everything back. Well, have a look at what you've done: the espalier tree is dead, one apple tree's had all its branches broken by snow and the young trees have all been killed off by the frost! You'll be the one paying for all of this damage!" It came as a real shock to me. Of course, I could not say that most of my plants at home were also dead, otherwise I would have had to pay for the damage inflicted everywhere.

Thank heavens there were customers I had not visited that year and so were spared my 'expert advice'. These customers were completely fine. Once I saw this I breathed a sigh of relief and decided to clear my head and forget what I had learnt. This put me back on the right track. Mrs Schuster-Bartl was completely right: the harsh pruning and the large amounts of fertiliser made the trees grow quickly. However, the branches could not lignify properly, so that they could not grow in the extreme temperatures we have in Lungau. I turned independent and healthy trees into dependent addicts and my harsh pruning only mutilated them. It was a great piece of luck that my practical experiences helped me to find my way back to the right and natural path.

My Method

According to my method of planting fruit trees you should leave all of the branches below the graft intact. This means that you should not cut them off! Also you do not dig a metre-deep hole, attach a tree guard, hammer in a stake or use chemical fertiliser. When I plant trees I just dig them in well and cover the area with mulch or with nearby stones. The layer of mulch holds in moisture, rots down and serves as fertiliser. The stones stabilise the tree by weighing down the roots. The stones 'sweat', in other words condensation collects underneath them, which is helpful for the newly planted tree. The stones also balance out the temperature. Finally, large numbers of worms can be found under them and these provide the tree with valuable and nutrient-rich worm casts. Many other important helpers like lizards, slow worms and ground beetles find a suitable habitat between the stones. Once I have planted the tree, I sow the seeds of soil-improving plants around it. Deep-rooted pioneer plants like lupins, sweet clover, lucerne and broom are particularly suitable. Their deep roots help to aerate the soil and prevent water from building up in the topsoil. Fruit trees are particularly sensitive to a build up of water. They become stunted, they no longer give the desired yield and are more susceptible to disease and pests. I am often asked why a particular tree does not grow properly. The reason is usually either that the location is wrong for the tree, whether it is too windy, warm, cold, wet or dry, or that the soil conditions are unfavourable. Compacted soil,

FRUIT TREES

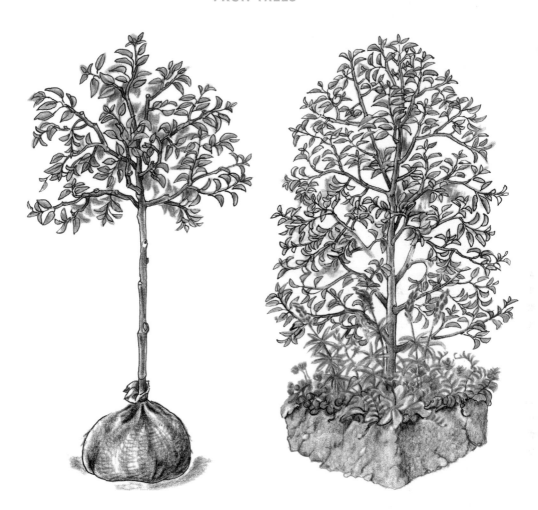

CONVENTIONAL METHOD:
A net is required to stabilise the root ball. Cutting causes the tree unnecessary stress and also requires work from people. The resilience of the tree is lost and the danger of damage from snowfall is higher!

MY METHOD:
A root ball dug out in a square plate shape: the root ball is well rooted and strengthened by the plant community. The tree can simply be replanted and will grow well. It should not be pruned back! The tree remains resilient, it can stabilise itself on slopes and the danger from snowfall is minimised.

above all, makes fruit trees difficult to manage. To improve the local conditions, I create microclimates such as suntraps, windbreaks or raised beds to give the tree the protection from the elements it needs. To improve the condition of the soil, I sow plant communities like those mentioned above. You must always observe trees closely, so that you get a feeling for when they are happy. In the

The branches sink down under the weight of fruit – this allows sunlight to reach the crown.

course of time, you will get a good eye for it and recognise straight away by the colour of the leaves and the bark whether a tree is in the right habitat or not. When I replant a tree, I dig the other plants up with it. This way I save myself the work of sowing another area. I dig the root ball up in a square plate shape, which is easy to put down on the ground or replant. The accompanying plants strengthen the root ball and protect the tree from drying out while it is being stored, transported and when it is re-establishing itself. As I do not require a net or mesh, I save myself the work that would otherwise be required when securing the root ball.

As I do not prune the trees, the branches retain their resilience. This means that they can support themselves on the ground when they are weighed down by fruit or snow. The trees can stabilise themselves and they are less likely to grow at an angle. They can adapt themselves to the terrain. When the branches are weighed down by fruit it allows sunlight to reach the crown. However, if I were to prune and ferti-lise the trees in the way that experts recommend, then they would put the excess energy into growing water sprouts, which results in a vicious circle. If I prune the fruit trees, they lose their resilience. The branches cannot sink when they are weighed down; instead they stick up rigidly in the air. At our kind of altitudes (up to 1,500m above sea level) these branches would not be able to withstand the weight of snow and would break. Too great a load of fruit would have the same result. Pruning the trees also creates wounds, which increase the risk of disease (fungal diseases, fire blight). It also causes unnecessary stress to the tree and requires a great deal of work from people. It is a shame that I did not plant fruit trees on the mountain pasture as a school boy. Twenty years ago, I began to go up to the mountain pasture and sow and plant it with a variety of different fruit trees. Today this fruit is particularly valuable, because the cherries ripen there in September. By this point the harvest is long over at lower altitudes where most cherries are grown, because the early varieties ripen at the end of June. At this altitude plums, pears and apples develop a very intense flavour, because of the harsh nights, so I can get a much higher price for them than fruit grown at lower altitudes. Distillery and vinegar specialists are very aware of these advantages, as well as those who are health conscious. It goes without saying that a culture of this kind contradicts all of the rules of conventional fruit-growing.

The side shoots and branches between the graft and the ground fulfil another important function. They provide protection against browsing. The twigs which are just above the ground are eaten by hares, the ones above are eaten by roe deer and the ones above those by red deer. The animals find food at the right height for them, so they do not damage the trunk. As the deer have free access to most of my polycultures on the Krameterhof and there are many tasty things for them to eat, a great deal of them do wander through, especially as the farm backs onto a large wild area. This means that my fruit trees require additional protection against browsing.

Fruit trees are planted on a newly created terrace. I sow various supporting plants around the trees, which improve the growing conditions for the tree as green manure crops and also provide grazing opportunities for deer as distraction plants.

In the picture, an apricot tree grows surrounded by sunflowers, Jerusalem artichokes, buckwheat, oilseed rape and scorpion weed, among others.

Protection Against Browsing

In general, the need for browsing drops when wild animals find enough food. For this reason, I always sow and plant everything I grow in large enough quantities so that the deer, birds, hares and mice all have something to eat. Nature is fertile enough to provide something for everyone. When humans become too miserly and want everything for themselves, a great battle against our many fellow creatures begins. To prevent large-scale damage from browsing, I introduce a large number of distraction plants, which the animals prefer to my valuable cultivated crops. Many plants like Jerusalem artichokes, various kinds of clover and buckwheat make good distraction plants and I also use them concurrently as green manure crops. A variety of types of fruit bush work well to distract deer and keep them away from the polycultures. To prevent stripping, I plant extra willow trees in the area and especially in front of slopes. The deer much prefer to strip these than the fruit trees, because they are softer and more elastic.

If there is a particularly high danger of browsing and stripping or the area available for cultivation is quite small, it is a good idea to provide additional protection for fruit trees. To do this I use a homemade remedy, which I paint on the trees or simply sprinkle over them. It is made of bone salve (the instructions to make this are in the 'Gardens' chapter), linseed oil, slaked lime, fine quartz sand and fresh cow dung. These ingredients are mixed to a spreadable consistency. If I want to sprinkle the salve on the plants, I add more linseed oil and use less sand. It is sprinkled the same way as holy water is in a church. The salve can also be applied with a brush or broom. The bone salve has an intense and long-lasting odour, which repels the deer. The smell gets into the bark and lasts for many years, in a way that is similar to mineral naphtha or beechwood tar, which can also be used instead of the bone salve. As neither naphtha nor beechwood tar have such an intense odour, it is a good idea to mix in singed hair, either pig bristles or cattle hair. To singe the hair, put it in a metal container and light a fire beneath it, so that the hair is singed by the heat. This produces a floury mass, which is then mixed into the salve. The smell of singed hair is only noticeable to humans for a short time, but it continues to be repellent to wild animals and they stay well away from the salve. The linseed oil is made from flax seeds and can also be bought in wholefood shops. The oil helps to bind the different ingredients of the salve together and makes sure that the salve adheres to the bark. The slaked lime is beneficial to the tree, because it emits heat. It also helps to combine the bone salve and other ingredients. The cow dung helps to bulk the salve out. It absorbs the other ingredients well, helps to repel animals and gives the salve a good consistency.

If an animal now tries to eat anything, the obnoxious odour will keep it away. If the smell of the salve begins to fade and the tree is under threat of being browsed, the quartz sand remains and causes an unpleasant sensation between the teeth. Once I observed a deer and her fawn from a raised hide as they tried to eat some of my young fruit trees. I had already sprinkled these trees with the salve. For the first couple of bites I could see no reaction. Then in the next bite there must have been a drop of my salve. The result: all of a sudden the deer began to act as if it were crazy. It gagged, threw its head from side to side, ran forward wildly and tried with all its might to wipe the taste from its mouth on the grass. The fawn reacted in the same way soon afterwards. I was in fits of laughter and had to get down from the raised hide quickly, because I could not hold myself up any longer. I was extremely encouraged by the effect of my salve. Obviously the deer did not like the taste of it at all. The salve has not let me down to this day.

Other possible ways to prevent damage from browsing and stripping are protective plants like wild roses, barberries, blackthorn or other similar thorny or prickly plants. The young shoots of these plants are grazed the most heavily, so they will become bushy and protect the fruit trees behind them.

Fruit Varieties

From my experiments, I have discovered that supposedly very demanding varieties – which, according to the experts, only thrive in warm climates and at low altitudes – can also adjust to high altitudes and give satisfactory yields. For example, Golden Delicious thrives here at 1,400m above sea level and produces large fruit, which store well. You should, therefore, not let yourself be dissuaded from cultivating so-called 'demanding varieties' at high altitudes. Naturally, they must be sheltered from the wind and it is important that they grow in climatically advantageous locations. Microclimates, which reflect heat from the sun using stones or bodies of water, for example, and protect the plants from weather extremes by using the effect of structures like hollows or windbreaks, are necessary for this. Under no circumstances should you turn to chemical fertilisers, because this will put the tree out of balance and it will not survive the winter. Fertilised trees grow faster and do not lignify as well as unfertilised ones. This means that they are not as frost resistant. According to expert opinion and literature, fruit growing ends at 1,000m above sea level in Lungau. Despite this, I cultivate a large variety of cultivated and wild fruit trees up to a height of 1,500m above sea level.

It is particularly important to investigate the different local varieties that grow in your area if you wish to grow fruit. These will probably be the best suited to your location. Hardy varieties I have had good experiences with can be found in the following lists. The ripening times given are the average for an altitude from 1,000m above sea level. They depend a great deal on the altitude and climate. The White Transparent apple ripens in the middle of August on the Krameterhof at 1,100m above sea level and can only be stored for a few days. As this variety ripens around St Bartholomew's Day on 24th August, it is known as 'Bartholomew's apple'. The fruit quickly becomes floury and, when harvested late, is barely suitable for pressing. At an altitude of 1,500m, on the other hand, White Transparent apples ripen in September, are much firmer and juicier, can be stored for over a month and are still excellent for pressing. The locations given are only general guidelines and should show where the best conditions for each variety can be found. The condition of poor soils can, however, be improved to a degree with green manure, by sowing supporting plants and creating microclimates. This can allow the majority of varieties to thrive on soils, which first appear to be quite unsuitable. So do not let yourself be discouraged from experimenting and gathering your own experiences with these fruit varieties.

Recommended Old Apple Varieties

Variety	Location and Characteristics	Ripening Time	Fruit Characteristics
Alkmene	Grows well on shady slopes, not susceptible to scab, mildew or frost, autumn apple	Mid-September	Dessert apple, can only be stored for a short time
Ananas Reinette	Demanding, small yield and greater chance of canker on wet soil, flowers early, winter apple	Mid-October, good to eat until March	Flavour similar to currants, dessert apple
Transparente de Croncels	Particularly suited to good, loose soils, wood and flowers not vulnerable to late frosts, recommended for high altitudes, good pollinator, flowers early and long, autumn apple	Beginning of September	Highly aromatic dessert apple, easily bruised, does not store for long (until October)
Baumann's Reinette	Prefers well-aerated soil, espalier tree, quite sensitive to frost, flowers early, winter apple	October	Yield is ample and early, firm fruit, dessert apple, good for drying, can be stored until April
Bohnapfel	Undemanding, prefers slightly damp soil, suited to harsh conditions (not sensitive to frost or wind), flowers late and long, winter apple	End of October, can be eaten from February to the end of May	Slightly acid, becomes juicy soon after harvest, good for pressing and cider, good storer
Boiken	Very undemanding in terms of soil and location, flowers and wood frost hardy, suited to harsh conditions, fruit is wind firm, winter apple	Middle to end of October	Sweet and fruity, given favourable conditions can be stored until May
Danziger Kantapfel	Undemanding, weather resistant, suited to high altitudes, good pollinator, flowers moderately early and long, not vulnerable to scab, winter apple	Beginning of October (can be eaten off the tree)	Juicy and aromatic fruit, good for juicing, can be stored until the end of January

Variety	Location and Characteristics	Ripening Time	Fruit Characteristics
Gravenstein	Particularly good on loamy soil, early variety (sensitive to frost), windfall, so should be sheltered from the wind, autumn apple	Mid-September	High-quality dessert and juicing apple, stores well (until December)
Jacques Lebel	Undemanding variety, flowers frost hardy, flowers moderately early and long, good for harsh conditions, but pay attention to wind-sheltering, winter apple	End of September until mid-October	Juicy, aromatic fruit, does not bruise easily, dessert apple, also well suited to drying and making cider, stores well (until December)
James Grieve	Well-aerated soil, also suited to cooler climates, flowers vulnerable to frost, autumn apple	Beginning of September	Aromatic and juicy, dessert apple, high yield
Jonathan	Suited to better quality soils, flowers moderately late, good pollinator, very susceptible to scab and mildew, winter apple	Beginning of October, good to eat until April	Rich in vitamin C, good dessert apple, stores well
Kaiser Wilhelm	Undemanding, flowers and wood frost hardy, well suited to high altitudes, fruit is wind firm, so it can be grown in windy areas, vigorous growth, flowers moderately early and long, winter apple	End of September until mid-October	Dessert, juicing and cider apple
Landsberger Reinette	Prefers wet soil, also suited to high altitudes, flowers long, not sensitive to the elements, suited to windy areas, winter apple	End of September until mid-October	Good dessert apple, good for drying, can be stored until January

Boiken apples: they continue to produce excellent fruit at high altitudes.

Variety	Location and Characteristics	Ripening Time	Fruit Characteristics
Maunzen	Undemanding, thrives in harsh conditions and at high altitudes, extremely frost hardy, flowers late, winter apple	End of October	Juicy, acid fruit, can be stored until March
Odenwälder	Undemanding, very hardy, suited to high altitudes, very frost and wood hardy, winter apple	Beginning of October	Aromatic and juicy, can be stored until December
Ontario	Prefers sunny areas, flowers moderately early and long, wood sensitive to frost, well suited to being an espalier tree, resistant to disease and pests, winter apple	End of October, can be eaten in January	Refreshing, juicy, acid, high in vitamin C, stores for a long time (until June)
Sheep's Nose	Prefers good soil, avoid extreme wet (danger of canker), flowers late, wood and flowers very frost hardy, suited to harsh conditions and high altitudes, winter apple	End of September until mid-October, good to eat until the end of February	Mildly aromatic, very good cider apple
Schmidtberger's Rote	Prefers wet, heavy soil, can be grown in sunny areas as well as in harsh conditions and at high altitudes, sensitive to overfertilising, winter apple	Good regular yield every other year in September	Juicy, acid fruit
Belle de Boskoop	Vigorous growth, fairly resistant to scab and canker; flowers early (sensitive to frost), winter apple	Harvest from the end of September to mid-October, can be eaten from December until February	Acid flavour, dessert and cooking apple, good for pressing
Stark Earliest	Hardy and undemanding variety, also ripens at high altitudes, early variety	August (generally before the White Transparent)	Aromatic, poor soil conditions lead to small fruit
White Transparent	Early variety, frost hardy (suited to harsh conditions and growing as an espalier), summer apple	August	Refreshing and juicy, can only be stored for a short time (around 14 days), dessert apple
Winter Rambo	Prefers fresh soil, flowers moderately late and long, flowers are resistant to late frost, very resistant to scab, winter apple	Beginning of October	Cooking and dessert apple, can be stored until January
Zabergäu Reinette	Moderately sensitive to frost, will grow on dry soil, low susceptibility to scab, flowers late and long, winter apple	Middle to end of October	Sweet and aromatic, dessert apple, large crop, can be stored until March

Alkmene in full bloom; this apple variety is well suited to shady conditions. In the picture it is being grown as an espalier on the western side of the Krameterhof.

Subira – a rare delicacy highly sought after in the production of schnapps.

Recommended Old Pear Varieties

As a rule, pears should not be harvested late, because they have a tendency to quickly become overripe and they cannot be stored any more. It is important to determine the right time to harvest.

Variety	Location and Characteristics	Ripening Time	Fruit Characteristics
Beurré Alexandre Lucas	Undemanding, deals well with frost, suited to high altitudes, good espalier tree, early winter pear	October	Sweet, fruity, refreshing dessert pear, stores well (until December)
Colorée de Juillet	Undemanding, suited to high altitudes, ripens very early	August	Cannot be stored, small, sweet fruit
Clapp's Favourite	Undemanding, must be sheltered from wind, not too dry, espalier tree, also grows in partial shade, summer pear	End of August	Large and juicy
Beurré Hardy	Undemanding, very vigorous growth, particularly good in wind-sheltered areas (premature windfall), flowers hardy, also suited to high altitudes, autumn pear	Middle to end of September	High-quality autumn pear
Comtesse de Paris	Prefers deep soil, wood is frost hardy, flowers sensitive to late frost, must be sheltered from wind, very good espalier tree	October	Tart and aromatic
Beurré Gris	Undemanding, can be grown in dry and windy areas, relatively small fruit, frost hardy	September	Very good for drying
Louise Bonne	Suited to well-aerated soil, flowers and wood sensitive to frost, wind firm, not in cold or wet areas, danger of scab, good espalier tree, autumn pear	September	Aromatic, sweet dessert pear, stores well (until November), very good for drying
Conference Pear	Undemanding, not sensitive to cold, avoid overly wet soil, autumn pear	September to mid-October	Very aromatic, good yield
Souvenir du Congrès	Undemanding, must be sheltered from wind, particularly suited to harsh conditions, espalier tree, flowers moderately early	Mid-September until the beginning of October	Very large fruit

Variety	Location and Characteristics	Ripening Time	Fruit Characteristics
Gros Blanquet	Very undemanding, hardy flowers, hardy variety, suited to harsh conditions, summer pear	End of July	Coarse-grained, aromatic, sweet dessert pear, also well suited to cooking
Doyenné Boussoch	Undemanding, hardy, suited to high altitudes, flowers and wood very resistant to frost, espalier tree, autumn pear	Mid-September	Tart, can only be stored for a limited time
Rote Pichelbirne	Particularly good on deep, wet soil, sensitive to frost, autumn pear	October	Juicy and sweet, good for cider and drying
Salzburger Pear	Suited to well-aerated soil, vulnerable to scab under unfavourable conditions, summer pear	End of August	Very aromatic dessert pear
Speckbirne	Suited to dry soil, flowers early, sensitive to frost	October to December	Particularly good for cider, also well suited to drying
Subira	Undemanding, thrives at high altitudes and in harsh climates	September	Exceptionally good for schnapps
Williams' Bon Chrétien	Does not need much sun, vulnerable to wind (windfall), still gives good yields in partial shade and at high altitudes, espalier tree, late summer pear	End of August	Particularly good flavour, very aromatic

Recommended Old Damson and Plum Varieties

Variety	Location and Characteristics	Ripening Time	Fruit Characteristics
Bühler Frühzwetsche	Undemanding, quite resistant to frost, disease and pests, ripens early	August	Juicy, but not very aromatic
Greengage	Undemanding, also grows on poor soil, wood and flowers quite sensitive to frost, grows in sheltered areas	September	Juicy, sweet fruit, cannot self-pollinate, very good for compote and jam
Quetsche	Suited to damp, warm areas and good soil (if it is too dry the yield and fruit will be small), sensitive to frost, but very good as a windbreak tree, self-pollinating	End of September to mid-October	Very sweet and aromatic, can be processed in many different ways

Variety	Location and Characteristics	Ripening Time	Fruit Characteristics
Kirke's	Undemanding, resistant to cold, well suited to harsh conditions and high altitudes	September	Large, sweet, juicy dessert plum, cannot self-pollinate
Czar	Prefers good wet soil in sheltered areas at high altitudes – ripens properly up to 1,400 m above sea level; hardy, but a little sensitive to frost	August	Juicy and mildly aromatic
Wangenheim's Early Plum	Undemanding, suited to high altitudes, resistant to frost, ripens fully in harsh conditions, self-pollinating	Middle to end of August (at low altitudes), mid-September (at high altitudes)	Juicy

Wild and Sour Cherries

Variety	Location and Characteristics	Ripening Time	Fruit Characteristics
Dönnisen's Gelbe Knorpelkirsche	Relatively undemanding, its colour means that it is rarely targeted by cherry fruit flies or birds	End of July	Firm, pleasantly aromatic yellow-gold fruit with clear juice
Große Prinzessin	Prefers good, deep soil, keep sheltered from the wind, resistant to cold, flowers moderately early and long	Mid-July	Aromatic, bright red with light coloured flesh
Bigarreau Noir	Prefers well-aerated, loamy and sandy soil, only slightly resistant to frost, good yields in windy conditions and at high altitudes	Mid-July	Very sweet, red-brown fruit
Hedelfinger Riesenkirsche	Adaptable, relatively resistant to frost	July	Juicy, dark brown-red fruit
Kassin's Frühe	Relatively resistant to frost, flowers early	June to July	Sweet to mild, red-brown fruit, well suited to juicing
Morello Cherry	Very adaptable, relatively undemanding, needs little sun, will thrive in wet places in partial shade (north-facing slopes, windy areas), wood is frost hardy, good pollinator, flowers very late	Beginning of August	Acid and tart, reddish-brown fruit, well suited to the production of juice, wine, compote and jam
Schneider's Späte Knorpelkirsche	Undemanding in terms of soil, quite sensitive to frost, flowers late	End of July to the beginning of August	Mild, reddish coloured fruit

Apricot and Peach Varieties

I particularly recommend ungrafted local varieties like the ungrafted vineyard peach, as it is less susceptible to leaf curl, a disease dreaded by peach growers.

Variety	Location and Characteristics	Ripening Time	Fruit Characteristics
Hungarian Best (Apricot)	Undemanding, will grow on poor soil, relatively resistant to cold, but vulnerable to late frost, flowers early, self-pollinating	End of July to August	Excellent for the production of jam and compote
Kernechter vom Vorgebirge (Peach)	Relatively undemanding, lives long, quite resistant to the elements	Mid-September	Sweet to tart flavour

Wangenheim's Early Plum

Propagating and Grafting

Cultivated fruit trees are not usually propagated using seeds, because the characteristics of the desired variety will not be passed on true through the seeds. This quality has been encouraged to the extent that many fruit cultivars can now only be pollinated by different varieties. This means that every flower, and therefore every fruit, can produce different seeds and consequently different characteristics. So, if you want to preserve the qualities of a certain variety, you must propagate it vegetatively. Grafting is fundamentally vegetative propagation. The shoot or 'scion' is not directly rooted, but instead joined to a rootstock. The practice of grafting was developed, because a tree with beautiful and aromatic fruit does not necessarily grow well. A grafted tree consists of at least two different plants with qualities which complement each other. This makes it possible to combine the positive growing characteristics of the rootstock with the qualities of the cultivated fruit to create a tree which grows healthily and produces good fruit.

Rootstock

For grafting to be successful both partners must 'get along' with each other. This means that only certain plants can be used as a rootstock. These are mostly members of the same species, although related species can also sometimes be used. As previously mentioned, the rootstock generally determines the growing characteristics of the grafted plant. It also has an effect on characteristics such as the plant's resistance to disease or frost. This means that for every variety of fruit there is a multitude of different rootstocks that can be chosen to match different criteria. Today, dwarf rootstocks are used for most fruit trees, because they keep the tree small and make it fruit earlier. An example of this would be the practice of grafting many pear varieties on quince as a rootstock, because it does not grow as vigorously and this slows down the growth of the pear. For my method of grafting, however, I prefer rootstocks from vigorous growing seedlings (from fruit grown from seeds) and wild varieties. Dwarf varieties used as rootstocks do not grow as vigorously. They do not develop strong root systems, which is one of the most important conditions for an independent tree. The weak roots mean that the trees frequently have to be tied to stakes so that they will not be knocked over by the wind or snow. They also cannot supply themselves as well with nutrients, which means that they have to rely on good soil or even fertiliser. They are more sensitive to drought and they are usually much more susceptible to disease and frost. For my requirements I need hardy and independent trees which can thrive on poor soil and in unfavourable locations. These needs are best fulfilled by rootstocks from vigorous and hardy seedlings. Their strong growth means that they will fruit a few years later and also grow taller, which makes harvesting the fruit a little more difficult. But I accept all of this happily. On the one hand, dwarf rootstocks would not grow so well on the farm. On

Cherry trees in full bloom on the site of an old spruce forest.

the other hand, the characteristics of these rootstocks and the cultivation they make possible would save me a great deal of work. If I take into consideration the amount of work I save in terms of maintenance by grafting onto vigorous rootstocks, the greater amount of work required to harvest them is put into perspective. A further point which speaks for using vigorous growing rootstocks is the fact that they live for much longer. So I can plant a fruit forest that will continue to provide food for the next generation.

Scion

You should use strong and sturdy perennial shoots for scions. Water sprouts are not suitable. The middle part of the shoot is used for grafting, which should have three to five buds on it. Scions should be cut during the dormant months in winter (January is best) and stored until they are grafted in the spring. It is best to store them in a cellar in wet sand. Scions can also be cut and used fresh for grafting. If you do this, it is best to use the shoots straight after they are cut. If I find a nice tree somewhere that I want to take a scion from and take home with me to enrich my orchard with, I have to stop the scion from drying out with a damp cloth. If the scions are cut in spring or summer, cut the leaves off and leave about 1cm on each leaf stalk.

Grafting

The aim of grafting is to bind the rootstock and the scion so that they grow together. It is necessary to achieve a good contact between the cambium layers of the rootstock and scion. The cambium is the cell layer between the bark and wood, which is responsible for growth by cell division. Only when these layers are well joined will the graft be successful. There are many different grafting techniques, which can be used on different parts of the tree and at different times. With a little skill and practice you can learn and carry out these techniques for yourself very easily. Neat work is particularly important. All cuts must be clean and you must not touch them, because this will contaminate the surface of the wound. For cutting you need a very sharp 'budding knife', which is only used for this purpose.

The author demonstrates the technique of cleft grafting.

- Whip and Tongue Grafting

For this method of grafting the rootstock and scion must be of the same thickness. I usually whip and tongue graft my young plants after the first or second year of growth in spring. I graft them at the root collar, which means that I cut the rootstock to around 10cm above the ground at an angle. The cut must be three to four centimetres long so that the rootstock and the scion make contact over a large area. It must be done in a single stroke, so that any unevenness is avoided. If the cut is not successful, it must be done all over again. I then cut a tongue in the rootstock (see the diagram 'Whip and Tongue Grafting'). The scion is also cut at an angle and cut with a tongue to fit the one in the rootstock. It is important to ensure that there is a bud on the opposite side of the cut. Both cut areas must fit together flush so that the cambium layers join together properly. Now the scion is slotted into the rootstock. The graft is then wrapped with raffia. The buds must remain uncovered so that they can still sprout. This binding is to improve the contact between the rootstock and the scion. To stop the graft from drying out or getting infected, it and any open cuts are painted with grafting wax. The buds must naturally not be covered.

- Cleft Grafting

A cleft graft is used when the rootstock is thicker than the scion. There are many different types of cleft graft, but the simplest is the bark graft. I generally carry out this type of grafting in May when the bark can easily be peeled back. The method is very simple. The stem of the rootstock is cut straight at the desired height and most of the twigs are removed. When doing this remember to leave one or two small twigs as nurse branches. They are important for supplying nutrients and also help to prevent sap from building up in the tree. The surface of the cut is then neatened with a specialist knife called a pruning knife, because clean cuts heal faster. A slit is now made in the rootstock without damaging the cambium and the bark is peeled back. The slit should be around 4cm long. The scion is cut at an angle as before. This cut should also be 4cm long. Again there should be a bud on the opposite side to the cut. To help improve the success of a graft, I also smooth off the edges of the area around the cut slightly (by roughly 1mm). Be sure to only smooth off the bark and not the cambium when doing this. This technique uncovers more cambium, which in turn makes it easier for the scion and rootstock to grow together. Now I push the scion into the gap underneath the bark. The bud level with the cut on the scion should be around the middle of the graft. Finally, the graft is bound with raffia and all cuts are painted with grafting wax. The buds should again remain uncovered. Depending on the thickness of the rootstock additional scions can be grafted. If the rootstock has a diameter of roughly 4cm or more a second scion should definitely be used. If the graft is successful, the nurse branches can be removed the next year.

WHIP AND TONGUE GRAFTING

Scion

The scion and rootstock are cut at an angle and a cut is made in each to make a tongue. They must fit together perfectly to ensure a good contact between the cambium layers.

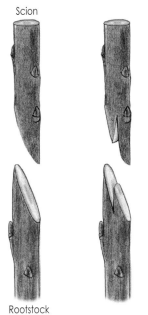

Rootstock

The graft is bound with raffia and painted with grafting wax. The buds must remain uncovered.

CLEFT GRAFTING

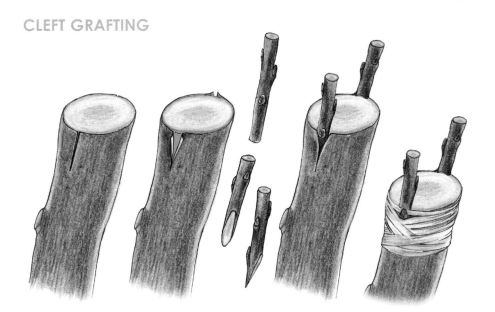

Vertical slits are made in the rootstock. The bark is carefully peeled back so that the scions can be inserted. The graft is bound with raffia and painted with grafting wax. The buds must remain uncovered.

BUD GRAFTING

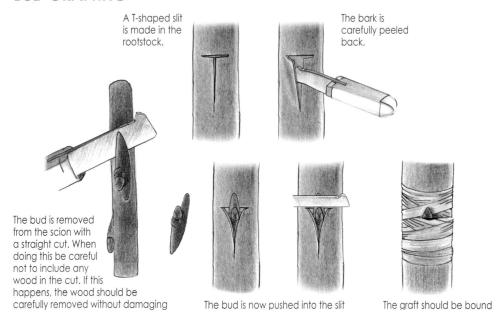

A T-shaped slit is made in the rootstock.

The bark is carefully peeled back.

The bud is removed from the scion with a straight cut. When doing this be careful not to include any wood in the cut. If this happens, the wood should be carefully removed without damaging the cambium.

The bud is now pushed into the slit and the bark sticking out over the bud should be trimmed flush with the T-cut.

The graft should be bound with raffia and painted with grafting wax whilst keeping the bud uncovered.

BRIDGE GRAFTING

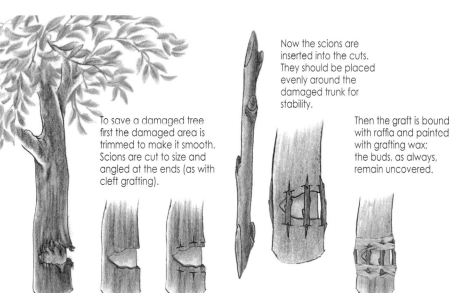

To save a damaged tree first the damaged area is trimmed to make it smooth. Scions are cut to size and angled at the ends (as with cleft grafting).

Now the scions are inserted into the cuts. They should be placed evenly around the damaged trunk for stability.

Then the graft is bound with raffia and painted with grafting wax; the buds, as always, remain uncovered.

T-shaped slits are made in the rootstock above and below the damaged area and the bark is peeled back slightly.

• Bud Grafting

Another method of grafting is bud grafting. In this case, only a single bud and not the entire scion is joined with the rootstock. A T-shaped cut is made in a smooth area of bark on the rootstock. Next the bark is peeled back slightly at the sides of the vertical cut. Then a well-grown bud is removed from a scion with a straight upward slice starting from the bottom (from the base in the direction of the tip of the shoot) in the shape of a shield. Be sure not to include any wood in the cut. Now insert the bud in the slit and push it in with the back of your knife. The bark overlapping the bud should be trimmed flush with the top of the T-shaped cut. Finally, the graft is bound with raffia and painted with grafting wax. The bud should remain uncovered.

You can use an active bud as well as a dormant one. An active bud should be budded in spring (May), and then it will sprout within the year. With a dormant bud you should bud in the summer (July or August). Dormant buds sprout the next year, hence their name. When grafting active buds you should use fresh scions. The leaves on these twigs are cut to short stalks. You will know whether the grafting has been successful if the leaf stalks fall off after around three weeks. Once the buds have joined with the rootstock properly and have sprouted, I cut the rootstock just above the graft and paint the cuts with grafting wax. When bud grafting it is also a good idea to use more than one bud. This increases the chance of success, because not every bud will necessarily sprout.

• Bridge Grafting

With the help of grafting I cannot only propagate fruit trees, but I can also save damaged trees. If a tree is heavily damaged, the flow of sap is interrupted or destroyed and the tree begins to dry out and die. If it survives, sooner or later the trunk will give way, because the damaged area begins to rot and the tree is destabilised. However, a tree in this situation can be saved relatively easily. I only have to join the area above and below with scions (preferably from the same tree), so that, in time, they can take over the transport and support functions for the trunk, in other words, the damaged area is bypassed. I begin by cleaning the wound and trimming all of the frayed areas. Now I graft scions above and below the wound underneath the bark. You should always use at least three scions.

This method can even help to save heavily damaged trees. Trees treated in this way are better protected from sources of damage (like browsing) in the future, because they are no longer easily accessible. The tree also has the resources to repair itself if it is damaged.

With grafting there are no limits to your fantasy: it is perfectly possible to graft a number of varieties onto a single tree. This is a great advantage when I only have space for one tree. As I have already mentioned, many fruit trees, especially apples and pears, cannot self-pollinate, so they must have pollinating varieties available. I can solve this problem by grafting on a branch from a pollinating variety. Having a number of varieties on one tree in a small garden helps to

In a fruit forest everyone is happy.

minimise the risk of crop failure. Different ripening times and different varieties of fruit make it possible not only to have a more varied harvest, but also creates a unique collection and increases the happiness gained from a single fruit tree. I believe that anything that is fun should be allowed in the garden. There are endless possibilities for experimentation. All you need is a little imagination.

Sowing a Fruit Forest

Using seedlings as rootstocks for my fruit trees is a very simple, economical and practically risk-free method of cultivating a lush fruit forest or orchard. I will now describe this method in greater detail.

Fruit trees generally prefer high-quality soil. I begin by preparing the area using soil-improving plants, so that my fruit trees will thrive. The role of green manure in the creation of humus is covered in 'Soil Fertility'. The amount of time this takes depends on the properties of the soil. The majority of acid soil on the Krameterhof, where spruce forests once grew, took around two years to improve to the point where I could grow fruit trees and other demanding plants without additional support. Green manure is not a one-off measure: it must play a continuous role in cultivation, because a fertile and healthy soil is the key to success. Once I have prepared the soil, it is necessary to loosen it for sowing. I graze my reliable workers, the pigs, there and they dig over and loosen the soil for me. This prepares the area for fruit trees and I can begin to sow the plants. The best and most economical source of seeds I have is in the form of pomace (the pulp left over from pressing fruit for juice or cider), although the material left over from the must when distilling schnapps will also work well if

The rowan (*Sorbus aucuparia*) is not only a beautiful tree, but its fruit is highly sought after in the production of schnapps.

the seeds can be separated after they have been heated. I leave this pomace to ferment for around four to five weeks and distribute it over the area. During the fermentation process the layers which naturally delay germination are broken down. This stratification greatly increases the seeds' chance for success. As the trees grow in their intended location from the start, they can best adapt themselves to the soil and climatic conditions. The diversity of plants lessens the chance of browsing from deer. Fencing the area off is never a bad idea, as this means it can also be used as a paddock. Once the trees have been growing for one to two years, they can be grafted. I only select the best trees for grafting, so I get the optimal plants for the location. Trees growing too close together can be replanted. This method is not only simple and economical, it is also particularly well suited to 'unfavourable areas', because my trees have had time to adjust to

were seedlings. This method should also
it requires very little effort and practically
spruce forest into a fruit forest using this
edlings in the area, because they can also
received very sweet fruit ideally suited to
re is a market for a great variety of kinds
ioned, and they should therefore not be
it in the form of fruit hedges, which can
e sensitive cultivated fruit trees. In this
ons. The fact that the hedges provide an
eful animals and insects is obviously also

very long and exhausting and took about
was a simple cart path and went through
d no end of interesting things there; a root
n a small tree which I would plant in my
of June, the end of school, I found a few
stones on my way back home. I could not
Although they were a good two metres tall,
ut digging, because their roots had found
joy, I carried them back home and wanted
I planted them. Instead of the praise I had
that it was a shame for the beautiful trees,
y would not take root at this time of year.
ittle garden (*Beißwurmhoanling*), dug them
overed the soil with leaves. I could not water
far away from the nearest source of water. I
trees growing. My mother had explained to
hen it was already far too late and they were
came upon the naive idea of removing all of
be stopping the trees from taking root. Then
I went to look at them every day in the hope
Several weeks passed until one of the trees
ise, produced new shoots. Once I discovered
ing her by the apron, I brought my mother to
believe my story, so I had to bring her, so that
was surprised and she asked me: "What did
hat luck!"
red me to develop my 'shock method'. It is
badly rooted trees without root balls to be
eady fully in leaf, in flower, or bearing fruit.

Fruit trees planted between raised beds using the 'shock method'.

I begin by laying the trees in the sun, so that the leaves dry out. Naturally, the roots should be covered, because they cannot tolerate sun. I use a wet jute sack to cover the roots. To make sure the leaves dry quickly, the trees must not be watered. The wet sacks will ensure that the roots do not dry out, but they will not provide enough water to supply the leaves. After about a day, the leaves will have dried out and the trees can be replanted. I do not soak the soil before planting or water the trees afterwards. The only protection they receive is a layer of mulch to keep the soil moist. I would never be able to water all of the trees on the Krameterhof, because it would take far too much time and energy. Trees planted using my method quickly develop new, fibrous roots, which supply the trees with nutrients and water again. They can survive the initial lean period, because they no longer have any leaves or fruit to support. If I were to plant a tree in full leaf and fruit instead and not water it, then all of its energy would be used to maintain the leaves. The roots would not get enough attention and the tree would grow badly, if at all. This tree could be compared to a cut flower: it is given plenty of water, yet it can barely support itself. Trees treated using my 'shock method' concentrate on taking root and do not produce shoots until they have the energy to do so. The trees are raised to be independent.

I have cultivated thousands of trees over the years using this method. I have bought remainder stock, which are often just chopped up or burnt, from tree nurseries at a very good price and planted them using my 'shock method'. In my experience, trees planted using this method grow best between raised beds. A large amount of moisture collects between the beds and the trees recover quickly.

After two to three years the trees have developed so well that I can dig around the root ball and replant or sell them. In this way my childhood experiences have provided me with a very good business.

Processing, Marketing and Selling

The diversity of an orchard provides a large range of processing and marketing opportunities. The dessert and cooking fruit is harvested and stored in a fruit or earth cellar. The fruit for juice, cider and vinegar as well as for making schnapps or for drying is sorted and processed. Fruit can also be made into jam or compote if there is a demand for it. Oil (from nuts) can also be pressed. As grafting and processing fruit takes a great deal of energy, the market situation should be carefully examined in advance. It is important to find out if there is a large enough market for the product and if you will get an adequate price for your efforts.

The success of your business also depends on the strength of its marketing. With a large farm like the Krameterhof, where we grow approximately 14,000 fruit trees of different varieties over an area of 45 hectares, it would be impossible to harvest all of the fruit we grow and process it. This is not only because the work required would be too extensive, but also because there are difficult-to-access areas, which make harvesting very difficult. In our case, the best use of these steep slopes is as a source of food for the pigs. The fruit trees grow just like any alder or spruce – the only difference is that every year they produce beautiful flowers and in autumn they yield fruit. They do not take any more or less time to maintain than any other tree. The fruit trees feed my livestock every year for a very long period of time without me having to do any work. Their flexibility is one further reason to try growing them.

Changing from commercial fruit growing to a permaculture system is really quite difficult. Usually commercial orchards are grafted on dwarf rootstocks and tied to espaliers. These dwarf rootstocks do not develop primary roots, because they no longer require them for support. A naturally growing tree on a seedling rootstock which is not supported by a stake naturally develops strong primary roots. It braces itself against the wind and develops into an independent tree that requires no further maintenance. In the espalier gardens of commercial fruit growing it is a good idea to make use of pigs. This means that you can leave all of the trees standing. They should no longer be fertilised or sprayed with pesticides and the yield should be used as a natural source of food for the pigs. For larger livestock like cattle and horses this is not advisable, because the many wires and narrow paths would present too many hazards. If the wires and stakes were removed, however, the orchard would be left in a sorry state. In this case, you must decide if you want to make a low-quality product, which for the most part will not be profitable. Changing to a permaculture system is not only difficult because of 'addicted' trees, but freeing yourself of the mind-set of commercial fruit growing also requires you to radically change the way you think.

Whilst I was providing a consultation on changing a farm over to using permaculture techniques in South Tyrol, an elderly farmer told me that he had been asked by the marketing cooperative to harvest his apples within a set period of

Fruit greatly enriches the diet of people and animals.

time. When I replied that the apples would still be green, he explained that the apples had to be green, otherwise they would be discarded as cheap pressing apples. He was not happy with the rates: the farmer said that he had only just received the invoice for the previous year and had to pay money back, because the storage costs were higher than the proceeds. I was surprised and asked him why he was still harvesting apples and delivering them. "Well, yes," he said, "but maybe things will go better this year." I told him that if I were him I would think of alternatives. "No, we can't do that here, we have contracts and we can't just get out of them. Also, what would people say?" Once he had admitted that he was making a loss and all that he was left with was work and the expense of chemical fertilisers and pesticides, he said, "Well, there's nothing you can do about that, that's just the way farming is. I can't do anything about it, you should explain that to my son."

We need creativity and courage to forge new paths. There are many ways to be a successful farmer. The higher yield of intensive farming, as the example shows, is no longer a guarantee of profitability – quite the contrary. The larger amount of work and the financial aid required often eat up the profits. How long will it take for farmers to free themselves of the shackles of cooperatives and make their way to independence?

4 Cultivating Mushrooms

Mushroom cultivation is, along with keeping livestock and growing plants, an important branch of our production on the Krameterhof. I began to work with mushrooms some time ago. Mushrooms were one of my most important sources of income in the 80s. In Lungau and the neighbouring areas I sold button, oyster, shiitake, and king stropharia mushrooms and many other kinds with great success. However, the atomic disaster in Chernobyl in 1986 suddenly changed the situation. Despite the fact that our mushrooms were obviously not contaminated, overnight it became impossible to sell them. This hard economical blow caused me a number of sleepless nights. In hindsight, it does, however, make clear what the consequences of extreme specialisation for a business can be. There will always be unforeseeable events, with unexpected outcomes, to which only the few with enough flexibility and versatility can adapt. Specialisation, on the other hand, creates only risk and dependence.

General

Fungi are not plants; instead they belong to a separate kingdom. They are one of the most important decomposers (saprotrophs) in the soil. They convert accumulated biomass into nutrients that plants can absorb. Without them the cycle of nature could not function. Anyone who wants to understand the way that fungi live and function must first understand their structure. The widespread opinion is that fungi only consist of the parts that can be seen above ground, in other words the cap and stem. In reality, a mushroom is nothing other than the fruiting body of a fungus and can be likened to an apple on an apple tree. The fungus consists of the much larger and mostly hidden mycelium, which is composed of thread-like cells (hyphae). If you compare the fungus to an apple tree, the mycelium corresponds to the trunk, branches and roots. Fungi, unlike plants, cannot produce organic material from inorganic material (mineral nutritive salts) for themselves. They do not contain chlorophyll and therefore cannot photosynthesise. This means that, like animals, they need nutrients from organic material (from the substrate), which are absorbed by the mycelium. Many fungi also develop mycorrhizae and form symbiotic associations with plants. The fungus' hyphae colonise the ends of the plant's roots, take over a part of the plant's nutrition and help the plant to absorb water and mineral nutrients. Nitrogen and phosphates are also made more easily accessible to the plant by the decomposing activity of the fungus. The fungus also profits from the symbiosis, because the plant supplies it with the products of photosynthesis (principally carbohydrates).

Many plants form symbiotic associations because of their inherent advantages. If a plant does not have access to its specific symbiotic partner, it will grow poorly. Frequently poor soil or unfavourable climatic conditions are blamed, whereas it is usually enough just to incorporate a little soil from the plant's natural environment. The previously stunted plant will now thrive, because it has access to its symbiotic partner. These basic concepts are needed to understand the way mushrooms are cultivated. The majority of cultivated mushrooms require a substrate of either wood, compost or straw.

Mushrooms that live in symbiosis, such as ceps (*Boletus edulis*) or chanterelles (*Cantharellus cibarius*), require their symbiotic partner in the form of forest trees in addition to a forest floor as a substrate.

If you understand what mushrooms need and the best ways in which to fulfil these needs, you will quickly achieve success in growing mushrooms. Mushroom cultivation does not require a large amount of space. It is possible to grow enough mushrooms for your own consumption on just a 2m² balcony. For farmers, mushroom cultivation can also make for a lucrative source of income with minimal costs and work. Before growing mushrooms on a large scale, however, it is important to gather experience and cultivate different kinds of mushrooms on different substrates. After some time spent experimenting, you will be able to use your experiences profitably.

Health Benefits

Mushrooms have long been recognised as not only a healthy food, but also as a form of medicine. This is why indigenous mushrooms like the honey fungus (*Armillaria mellea*) have been used as a laxative for centuries. The giant puffball (*Calvatia gigantea*) and agarikon (*Laricifomes officinalis*) mushrooms were also used to staunch bleeding. However, this old knowledge has mostly been lost.

Now that Asian medicine with its natural remedies has become so popular, mushrooms are once again being considered as a medicine. One of the most interesting East Asian medicinal mushrooms is the shiitake mushroom (*Lentinula edodes*). This mushroom is not only a very popular delicacy because of its excellent flavour, but its healing effect is astonishing. Its ability to lower cholesterol has already been demonstrated in medical studies; it is also effective against colds and strengthens the immune system. Moreover, it has been scientifically confirmed that the shiitake mushroom has a positive effect in the treatment of cancer. Although probably the most interesting thing about this medicinal mushroom is that it can be cultivated almost anywhere and with very little effort. Shiitake cultures grow on sycamore logs on the Krameterhof up to a height of 1,500m above sea level.

SYMBIOSIS BETWEEN A TREE AND FUNGUS

using the example of a birch bolete (*Leccinum scabrum*)

Advantage for the fungus: the tree supplies its symbiotic partner, the fungus, with carbohydrates.

Birch – producer/ photosynthesiser

Mycorrhiza

Advantage for the tree: decomposition and minerals – the fungus provides water and mineral nutrients. The breaking down of humus greatly enriches the soil with nitrogen and phosphates.

Leaf fall

Birch bolete – saprotroph/ decomposer

There are also a number of other medicinal mushrooms, which can easily be grown on the Krameterhof. For example, the Judas' ear fungus (*Auricularia auricula-judae*), which can be used to treat nausea or to lower blood pressure, or the lingzhi mushroom (*Ganoderma lucidum*), which is used to treat sleeping disorders and strengthen the immune system.

Mushrooms are also a healthy food. Their high fibre content is known to aid digestion. They are also low in calories and rich in vitamins and minerals, which means they are often used in diets. Mushrooms are best eaten fresh, but they retain much of their flavour and nutrients when they are dried. They also make very good tea. Mushroom tea is very good at preventing and treating illnesses and for detoxification.

The Basics of Mushroom Cultivation

Most mushrooms need wood, straw or compost as a substrate. Mushrooms that grow on wood or straw are by far the easiest to cultivate, because wood and straw are already complete substrates. Compost mushrooms like button and field mushrooms (*Agaricus spp.*) or shaggy ink caps (*Coprinus comatus*) are, conversely, much more difficult to cultivate, because they need a specific mixture of compost made from straw or dung (usually horse manure). Producing the substrate requires specialist knowledge and is usually too laborious to be used on a small scale. This is why I limit myself to cultivating mushrooms that can be grown on wood or straw. Most of the particularly tasty and healthy mushrooms belong to this group anyway. You will not need any specialist knowledge to be able to cultivate these mushrooms successfully.

Only uncontaminated raw materials should be used to grow culinary and medicinal mushrooms. Mushrooms can absorb harmful substances and retain them. For this reason you should be especially careful when using straw or compost as a substrate. If the straw or manure has been produced conventionally it is, in my opinion, unsuitable for cultivating organic mushrooms. Even wood can contain harmful substances. Trees near busy roads, motorways or industrial areas usually contain large amounts of heavy metals. These harmful substances accumulate in the bark and make the logs unusable for mushroom cultivation.

Growing Mushrooms on Wood

The fact that most mushrooms that grow on wood are cultivated in almost exactly the same way makes growing them much easier. Any differences are usually just limited to slight preferences in terms of log size, type of wood, temperature and humidity. If possible the mushroom culture should be started in spring, because this gives the mycelium enough time to colonise the wood and remain safe from frost damage. The mycelium can still grow at low temperatures, but quick colonisation occurs best at temperatures around 20°C. If the fungus has the chance to grow deep into the wood through the summer, then low temperatures and frost will not be able to harm it. There are, however, different temperature preferences for producing fruiting bodies. For instance, shiitake mushrooms prefer temperatures between around +10°C and +25°C. Enoki mushrooms (*Flammulina velutipes*), on the other hand, produce fruit at low temperatures in late autumn. This is why it is always important to cultivate different kinds of mushroom, so that you will have a yield over an extended period of time. Many different kinds of mushroom are suited to being grown on logs. Some of the most common ones are listed below.

Mushrooms for Growing on Wood

All of the mushrooms listed grow on the Krameterhof on hardwood. The oyster mushroom varieties listed (*Pleurotus sp.*) can also be cultivated on straw. Fruiting body production occurs in most of these varieties from a temperature of at least 10°C. Only the enoki mushroom fruits from temperatures of 2°C. The black poplar mushroom, on the other hand, grows fruiting bodies from around 15°C. All of the oyster varieties, especially the king oyster mushroom, grow best when the humidity is high.

- Shiitake

The shiitake mushroom (*Lentinula edodes*) is one of the most interesting of the culinary and medicinal mushrooms with its excellent flavour and many proven healing properties (cf. 'Health Benefits'). It can also be cultivated on narrow logs or branches. Shiitake mushrooms can be eaten raw, or they can be used to make tea.

- Oyster Mushroom

Oyster mushrooms (*Pleurotus ostreatus*) are excellent culinary mushrooms and are very easy to cultivate. Although they are not particular about the kind of wood they are cultivated on, they grow especially well on beach, maple and elm.

- King Oyster Mushroom

King oyster mushrooms (*Pleurotus eryngii*) are very popular because of their excellent flavour. Their thick fleshy stems are very versatile. They can be grown like oyster mushrooms.

- Golden Oyster Mushroom

Golden oyster mushrooms (*Pleurotus citrinopileatus*) can be recognised by the way they grow in large yellow clusters. They are excellent culinary mushrooms. They can also be cultivated in a similar way to other oyster mushrooms.

- ## Sheathed Woodtuft

These small mushrooms (*Kuehneromyces mutabilis*) grow in clusters. They have a very intense flavour which means they are used predominantly as culinary mushrooms in sauces and soups. Sheathed woodtuft mushrooms are particularly undemanding in terms of cultivation.

- ## Nameko

This excellent culinary mushroom (*Pholiota nameko*) is also known as the Japanese sheathed woodtuft. Its requirements are similar to the sheathed woodtuft.

- ## Enoki

The enoki mushroom (*Flammulina velutipes*) is also known as the winter mushroom, because it bears fruit in autumn and winter. It is used as a culinary mushroom and for flavouring sauces and soups.

- ## Black Poplar Mushroom

Like the sheathed woodtuft, it is very aromatic and is mostly used for adding flavour to dishes. The black poplar mushroom (*Agrocybe aegerita*) requires much higher temperatures. It prefers softwood (poplar and willow).

- ## Judas' Ear

The Judas' ear fungus (*Auricularia auricula-judae*) is popularly used in Asian cuisine. It is also a popular medicinal fungus. Judas' ear prefers to be cultivated on elder, but it also grows well on other kinds of wood.

Substrate

For wood cultures I generally use whole logs as a substrate. It is also possible to use a substrate mixture of sawdust and other plant material, although using a mixed substrate involves a little more work and is a bit more risky. For this reason, growing cultures on natural wood is much better for beginners. In addition, the wood is used in its natural form, which saves the work of processing it, which can be considerable. Logs have the additional advantage that they take up far less space than other substrates. They also improve the look of any garden. It is important that only hardwood is used for the mushroom species specified above, although in my experience out of all the hardwoods, wood from stone fruit is the least suitable for mushroom cultivation. The duration and yield of the crop differs greatly depending on whether hardwood (beech, oak etc.) or softwood (poplar, willow, alder, birch etc.) is used. Mushrooms colonise softwood much more quickly, which leads to an earlier yield. However, softwood logs also decompose more quickly, so the yield is much more short-lived. Cultures on softwood can grow through and give a yield in just six to twelve months. Hardwood usually requires twice that time, but the yield lasts substantially longer. Naturally, the duration and size of the yield also depends on the size of the log and the length of the growing season. For instance, I have cultures growing on hardwood at a height of 1,500m that have been cropping for over ten years. As a rule, under good conditions you can expect a total yield of 20 to 30 percent of the weight of the wood. This can be very profitable, because financially low quality wood like firewood can be used for mushroom cultivation. In the long run, hardwood generally gives higher yields than softwood.

One of the most important factors for success when growing mushrooms in this way is using fresh and healthy wood. It must under no circumstances have been previously colonised by other fungi. They will displace the cultivated mushrooms and the crop will fail. This is why I use freshly cut wood whenever possible. It is also advisable to cut off a slice of wood from each end of the log before inoculation. This will reduce the danger of colonisation by other fungi. Wood that has been stored for more than half a year is probably no longer suitable. Tree stumps are an exception: they should be inoculated once they have stopped sprouting. These stumps can still reject mycelia. You can recognise healthy wood by it not having any dark or rotten areas. The cut surface must be light in colour and firm. The wood must contain enough moisture for the mycelium to grow through properly. Freshly cut logs have the best levels of moisture; if you use older logs, you will need to soak them for a while to reach the required moisture levels.

As hardwood can be used in practically any form, there is almost no limit to the types of culture you can try. This means that, for instance, logs used as short term slope supports can also be used to grow mushrooms on at the same time. Tree stumps in the garden can easily be broken down with the help of mushrooms. The mushrooms will make the garden look more pleasant, as well as giving a good crop. As shorter logs are much easier to manage, I generally use logs with a length from half a metre to a metre and a diameter of at least 20cm for growing mushrooms. These logs also take less time to colonise and therefore give an earlier yield.

Mushroom Spawn

In order to inoculate the log, in other words to introduce the fungus into the log, you will need healthy mushroom spawn. Mushroom spawn is nothing other than mycelium, because mushrooms reproduce vegetatively (asexually) as a rule. Inoculation using spores (sexual reproduction) is rare, because the probability of failure is too great. Although the sheathed woodtuft mushroom (*Kuehneromyces mutabilis*) and enoki mushroom (*Flammulina velutipes*) are an exception to this; they are easy to propagate by placing ripe caps on the ends of moist logs and tree stumps (they prefer softwood: poplar and willow). If the spores find the right conditions, they will germinate and the fungus will colonise the whole log. This method does, however, take much longer than vegetative propagation. Naturally, all mushrooms can be propagated using spores. As it does not take much time or cost much money, you should definitely try it. However, if you only have a few logs available, inoculation is a much safer choice. Mushroom spawn is usually available from specialist mushroom shops as spawn plugs or grain spawn. Spawn plugs consist of wooden plugs or dowels that are inoculated with the appropriate mycelium, whereas grain spawn consists of mycelium growing on grain. It is important for the spawn to be healthy. The mycelium of cultivated mushrooms is white. If there is a change in colour, then this is a sign of mould. Also, a musty smell means that the spawn is either contaminated or has expired. The spawn should be used as soon as possible, because it can only be stored for a limited time.

• Propagating Spawn Yourself

You can create and propagate mushroom spawn yourself with a little practice. Of course, propagating with spawn is a learning process, so you might not get the results you want on your first attempt. The majority of failures stem from careless work and the mould associated with it. Spawn plugs are not as susceptible to mould as grain spawn or substrate spawn, which makes them a suitable method of propagation for beginners. If you want to make more spawn plugs, all you need are plugs or dowels. They must be soaked and boiled to reach the required moisture levels and to rule out contamination by rival fungi. The boiled plugs then go into a clean plastic bag. Once they have cooled, add a few inoculated spawn plugs or some grain spawn. The bag is then turned over so that the opening is at the bottom. This way the right amount of air can get in without the danger of contamination becoming too great. After about a month, the plugs will be fully colonised and ready for inoculation. It is also possible to wrap boiled dowels in a piece of clean cotton cloth with some mushroom spawn. The wrapped up dowels can then be placed in a flowerpot to keep them moist; storing the substrate on soil keeps it wet, while excess water can drain away. This prevents water from building up and the dowels should be fully colonised in roughly one month.

Creating and Maintaining the Culture

To inoculate a log with mushroom spawn, you will need either to drill a hole or cut a section out of it. The method used will depend on the type of spawn available. With spawn plugs, a hole is drilled into the log and the plug is inserted. It is important to achieve good contact between the plug and the wood, so the holes should be only a little larger than the plugs. When you are inoculating you should use plenty of spawn. The plugs should be distributed evenly around the log to ensure that it is colonised quickly. It is also a good idea to seal the holes again after inoculation; you can do this by pushing a fresh piece of branch into the hole and cutting the excess off. Now the fungus is protected and it can colonise the log quickly.

Another method is 'notched inoculation'. This is where a saw is used to make notches in the log in either one or two places depending on its length. The notches should have a depth of over half the diameter of the log, however, it must retain its structural integrity, which is why I make the notches using a chainsaw (see photo). These notches are then filled with grain or substrate spawn. Next, I cover the notches with plastic sheeting or adhesive tape. This is necessary to prevent the mycelium from drying out or from being contaminated by mould. Covering the notches also prevents the fungus from being eaten. This is a good idea, because slugs, snails, birds and mice like to eat mushroom spawn as well as cereal grain.

To reiterate, as mushrooms require a steady temperature and moisture level for good growth, I place the culture in an area with plenty of shade. The inoculated logs are stored closely together. To help prevent them from drying out I cover them with leaves and jute bags. The best temperature for mycelial growth for these mushrooms is around 20°C; at lower temperatures the mycelium takes longer to colonise the log. However, temperatures of over 30°C should be avoided, because they can kill off the mycelium. It is best to start the mushroom culture

Notching the logs

Notched logs

Covered inoculation area

In the inoculation area the mycelium is growing well through the wood.

in spring or early summer, because the mycelium needs two to three months to work its way far enough into the wood. After this time, the fungus is safe from frosts. The duration of the colonisation phase depends on the temperature, moisture and the size and type of wood (hardwood or softwood). Although, as a rule, it takes between six and twelve months. After just a few weeks, you can find out if the inoculation has been a success. If the white mycelium is establishing itself in the inoculated area, then the mushroom culture is healthy. As soon as the mycelium can be seen in the notch, the fungus has colonised the log.

Once the log is colonised, I sink it into the ground by a third of its length in its direction of growth (the thicker end should be at the bottom). The space between the logs must be large enough so that the mushrooms growing on the outsides of the logs can be harvested properly. Sinking the log into the ground is very important for the success of the culture and means that it will need little maintenance. This also allows the fungus to obtain moisture and nutrients from the soil; helping to prevent the log from drying out and reducing the amount of

Oyster mushroom (*Pleurotus ostreatus*)

GROWING MUSHROOMS ON WOOD

Fruit tree provides shade

Storage:
A layer of leaves and jute bags provides optimal conditions during the sensitive colonisation phase. The logs can, however, be sunk into the ground in their final intended location immediately after they have been inoculated (as with logs that have already been colonised), but the colonisation phase may take longer.

Logs during the colonisation phase

Culture: The logs are now placed further apart, which gives the fruiting bodies more space to grow. The colonised logs are sunk into the ground by a third of their length. This allows the fungus to take additional nutrients and moisture from the soil.

Mycelium growing through into the soil

work required to maintain the culture to a minimum. If you pull one of these logs out of the soil after a few weeks, you will discover that the mycelium has already grown through into the soil. Fruiting bodies have even been known to spring up around the log. However, the main source of nutrients will still come from the log itself.

If humidity levels are high enough and the temperature is correct, I can soon expect a yield. The culture usually crops a number of times a year. Mushrooms generally appear near the inoculation area and on the outside of the log. If the mushroom culture is in the correct environment, it will need practically no further care apart from maintenance of moisture levels. It can remain in the same place and will not require any special protection over the winter.

Tips

If, after a few years, there has still not been a crop and the mycelium has not grown through well, this means that the conditions for fruiting body growth are not optimal. 'Dormant' logs can, however, be activated by soaking them for a few hours and striking them with a mallet or rock. The moisture and the shaking stimulate fruiting body growth. This method generally has the desired effect. I discovered this phenomenon when I first began to cultivate mushrooms and some of my cultures were not successful. After waiting for a while I decided to dispose of them. I took what I supposed to be useless logs on my tractor to one of the wetlands and tipped them all in the shallow water near the bank. I wanted the logs to at least provide hiding places for young fish and crayfish. They also made the shallows of the wetland more attractive. When I came back to the microclimate after a few weeks, almost all of the logs were covered in oyster mushrooms. I could scarcely believe my eyes and tried to find the reason for the 'resurrection' of my cultures. What happened? The logs were most likely dormant from lack of moisture. The bumpy ride on the tractor and being tipped into the pond not only shook them thoroughly, but also saturated them with water. In addition, the hollow around the wetland is naturally much more humid as a result of evaporation. All of these factors worked together and led to this unexpected success.

Another common problem is browsing. Humans are not the only creatures who like to eat mushrooms; animals do as well. On the Krameterhof the non-indigenous Spanish slug (*Arion vulgaris*) particularly likes to eat mushrooms. Slugs and snails can do a great deal of damage. This damage often remains unnoticed, because there are never any mushrooms left. On the Krameterhof we do not have any problems with these pests, because we have plenty of help from pigs, ducks and toads. These helpers reduce the number of slugs and snails to a harmless number. If you do not have such helpers available, there are always long-established household remedies. One of these remedies is to put down a protective ring around the culture. Make the ring from a mixture of wood ash, sawdust and slaked lime. It is important to keep it dry at all times, so that it forms an unsurpassable barrier. A further possibility is a line of fresh grass cuttings. This line must be kept wet. It will attract slugs and snails and they will lay their eggs in this inviting environment. After a few days, turn the line over and leave the eggs in the sun. This method can drastically reduce the number of offspring, because the eggs dry out and are sensitive to UV light. Further information on slugs and snails can be found in the section 'Helpers in the Garden and Regulating "Pests"'.

Growing Mushrooms on Straw

These days straw is often a waste product. In areas where cereals are grown you can all too often see it being left to rot in enormous piles in fields. However, this 'surplus' biomass can be made good use of. Straw can, for example, be used as a building material for mudbrick buildings. It can also be used as mulch and it is an ideal substrate for mushroom cultivation. Many kilograms of mushrooms can be grown on one small straw bale with very little effort. The mushrooms also help to turn the slowly rotting straw into valuable humus in just a short period of time. So many of these so-called waste products are untapped resources, which could be made use of in the future. In a functioning agricultural system there is no waste, everything can be brought back into the cycle of nature. Sustainability is the highest priority here. With minimal work, growing mushrooms on straw can be made into a lucrative additional source of income for farmers. They can also be grown easily in any small garden for small-scale consumption.

Mushrooms for Growing on Straw

All of the previously mentioned oyster mushrooms (*Pleurotus sp.*) are suitable for growing on straw. Another variety of mushroom that can be cultivated very well on straw is the king stropharia mushroom (*Stropharia rugosoannulata*). From a distance this mushroom looks similar to a cep. It is an excellent culinary mushroom and can be grown without any difficulty. It requires temperatures of over 10°C for growing fruiting bodies. Its requirements for humidity are, however, much lower than those of oyster mushrooms, which makes king stropharia mushrooms a little less work to maintain.

Substrate

In principle, any kind of straw can be used for cultivation. There are, however, a few basic requirements that the substrate must fulfil. It is particularly important that the substrate is healthy and in good condition. As I have previously mentioned with regard to growing on wood, cultivated mushrooms are very sensitive to competition. Straw that has already been contaminated by other fungi is unsuitable and cannot be used for cultivation. Healthy straw can be recognised immediately without any specialist knowledge; it should be a natural golden

THE CYCLE OF GROWING MUSHROOMS ON STRAW
– AN EXAMPLE OF SUSTAINABLE FARMING

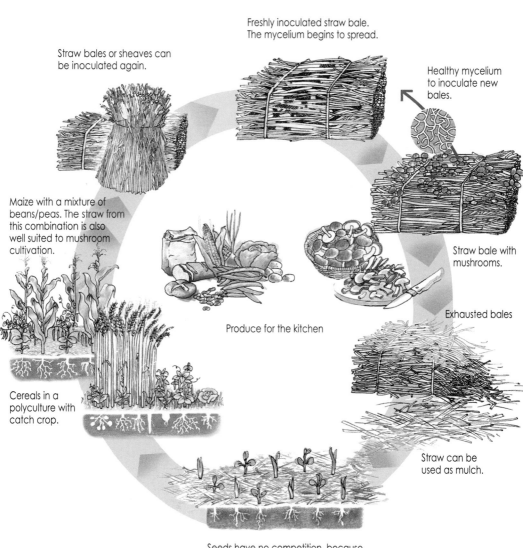

Freshly inoculated straw bale.
The mycelium begins to spread.

Straw bales or sheaves can
be inoculated again.

Healthy mycelium
to inoculate new
bales.

Maize with a mixture of
beans/peas. The straw from
this combination is also
well suited to mushroom
cultivation.

Straw bale with
mushrooms.

Produce for the kitchen

Exhausted bales

Cereals in a
polyculture with
catch crop.

Straw can be
used as mulch.

Seeds have no competition, because
the other seedlings rot underneath
the layer of mulch.

yellow colour, must not have any spots of mould or dark areas and must not smell musty.

Small compressed straw bales are the easiest to work with, because they can still be transported easily when they are wet. For me it is essential that the straw comes from organic farms. Conventionally grown straw is not suitable as far as I am concerned, because it might have been treated with or contaminated with herbicides, insecticides, fungicides, chemical fertilisers and other chemicals such as growth regulators. For this reason, you should use straw that has been organically grown.

In addition, straw provides the main source of nutrients for the fungus, which will absorb and retain any harmful substances present. This is yet another reason why conventionally grown straw is not suitable for producing food. Finally, the chances for success for a culture of this kind are much lower with non-organic straw, because it is usually treated with fungicides. Fungicides are used to fight fungi. Logically, a substrate treated in this way is not ideal for mushroom cultivation. The slightly higher price for organically grown straw is worth it in any case.

Since these mushrooms, as I have already mentioned, break down organic materials (saprotrophs), straw is not the only available option. They can be grown just as well on reeds, shredded garden waste or many other substrates. However, these altrenative substrates are not as easy to work with and it will require a little experimentation from time to time to get the correct composition for a successful crop.

Mushroom Spawn

Straw cultures can also be inoculated using either substrate spawn or plug spawn. The one you choose is, in the end, a matter of personal preference. The requirements in terms of spawn quality are the same as those for mushrooms being grown on wood. Spawn propagation also works in the same way. There is, however, another method to prolong the life of a straw culture. To do this I remove some mycelium from a healthy bale that has been well permeated and introduce it into a new bale. The mycelium will usually spread from a permeated to a fresh bale with only a brief period of contact. This way I save myself not only the effort of buying new mushroom spawn every year, but also the task of inoculation.

Creating and Maintaining the Culture

One of the most important criteria for the success of a culture is for the substrate to have high enough moisture levels, so the initially dry straw must be thoroughly soaked. To do this I leave the bales for a number of days in a container filled with water. Submerging the bales fully not only helps me to achieve the required level of moisture, but under these conditions the bales also begin to ferment slightly. This makes it easier for the mycelia to colonise them. Then I put the bales out for a day or so to allow the excess water to drain away. Once this is done, the straw fulfils all the requirements for inoculation. Next the bales are brought to their final location. As with log cultures, this should be a place with plenty of shade.

On the Krameterhof I use my wetlands to soak straw bales. I then place the mushroom cultures on the banks. This saves long journeys and the high level of evaporation from the nearby water ensures an optimum level of humidity. In addition, this makes it easy to water the culture during dry spells.

When positioning the bales, you should make sure that there is enough space between them, because mushrooms will grow over the entire bale. Next it is time to inoculate the bales. If you are using substrate spawn, you should make a number of holes in the bales with a stick. These holes should reach at least as far as the middle of the bale. Now fill the holes with spawn and close them again by pushing the straw together. If you are using plug or dowel spawn, you should push them into the bales in an even distribution. The plugs should also be pushed in as far as the middle of the bale.

With both methods it is important to use plenty of spawn, because a large amount of evenly distributed spawn is needed for quick colonisation. The more rapidly the bales are colonised, the less likely it is that other fungi will contaminate the culture. In a single straw bale, I inoculate in approximately eight to ten places. As I have already mentioned, the length of time it takes for colonisation depends a great deal on the temperature. The optimal temperature for mycelial growth for mushrooms growing on straw is a little over 20°C. If the cultures are inoculated in spring or early summer, you can expect a yield in roughly three months. Cultures inoculated in autumn, on the other hand, will not produce a crop until the next spring. Well-permeated straw bales are not normally sensitive to frost. You can easily recognise a well-permeated bale by the white pleasant-smelling mycelium of the cultivated mushroom that has grown through the straw. You should also make sure to leave cultures that are inoculated in autumn enough time to colonise.

Straw cultures require little additional maintenance. You only need to check the bales regularly to make sure that they are moist enough. It is not a problem if the straw dries out for a few centimetres around the outside of the bale as long as the middle is still wet. Often people well-meaningly overwater mushrooms. Although the mycelium requires moisture, it is still susceptible to ongoing dampness. If the bales are going to be left outside exposed to the elements, they can be covered during heavy or long periods of rain. I leave all my straw cultures outside throughout the year, however, without any additional protection against the weather. The bales can, of course, be covered with brushwood to offer protection over the winter months.

The yields usually occur in phases, provided that the humidity and temperature levels are high enough for the type of mushroom in question. For this reason, the culture's yield depends not only on the size of the substrate, but also on environmental factors. Straw cultures have a life span of between one

ENVIRONMENT OF MUSHROOMS
IN A POLYCULTURE

Fruit trees to provide shade

Mushrooms grown
on logs

Mushrooms grown
on straw

Evaporation ensures optimal
humidity for mushroom growth.

The wetland provides the
system with water and can be
used to prepare straw bales.

King stropharia mushrooms on a straw bale in the second year.

and two years. Afterwards, the bales are worn out and can be used as fertiliser or mulch.

Tips

Worn out bales make excellent mulch. They can also be used in the garden as organic fertiliser. When doing this it is not uncommon for mushrooms to suddenly appear in the vegetable patches.

Cultivating Wild Mushrooms

Over the years, I have made numerous attempts to propagate and create cultures of ceps (*Boletus edulis*), chanterelles (*Cantharellus cibarius*), birch boletes (*Leccinum scabrum*) and a number of other mushroom varieties with great success. These mushrooms live in symbiosis with forest trees. In order to cultivate them, you must provide them with the correct location and symbiotic partner. On the Krameterhof this always occurs in a mixed culture, because the various tree species complement each other not only from an ecological point of view, but also in terms of mushroom cultivation. Spruce trees (*Picea abies*) are, in fact, an important symbiotic partner for ceps and chanterelles on the Krameterhof, however there are usually very few mushrooms to be found in spruce monocultures. A mixed forest is not only more stable; the humus composition is also better. Pine needles rot very slowly on their own and, in the long run, contribute to the acidification of the soil. Even the balance of water in mixed forests is better and this, in turn, plays a large role in good mushroom growth.

There are many factors that are responsible for a successful culture. Fulfilling them requires a very close observation of natural cycles. This means that cultivating these mushrooms is still not commercially viable to this day.

I would like to explain my method of cultivating wild mushrooms using birch boletes as an example. First I create the correct growing environment. To do this I might plant young birch trees on a terrace. Afterwards, I introduce the mycelium into the soil between the young trees. Obtaining the mycelium in the first place is somewhat more difficult. To do this I need an area where birch boletes are already growing. To obtain fresh mycelium, I use a substrate mixture of my own and spread it around the growing fruiting bodies. This substrate consists of forest soil and sawdust soaked in liquid manure made from different symbiotic plants.

I spread the substrate when the first mushrooms appear and leave it there for the length of the growing season. I check to see if the mycelium has grown through by carefully examining the substrate for fine white mycelial threads. The substrate can then be introduced at different depths in the soil around the birch trees that were planted in the spring or around birch stumps that are still living and producing shoots. It is also possible to carefully place the substrate directly in the hole when planting trees. If this is done correctly and the conditions are favourable, fruiting bodies can grow as soon as the next year.

Another very good and simple way to propagate mushrooms is to collect ripe fruiting bodies (with ripe spores) and to put them in a loose-weave bag (jute or onion bag). The bag can then be hung up to dry. It is important that it hangs in a well-ventilated area, so that it can dry out properly. Once the mushrooms have dried out, I walk around the area striking the bag lightly with a stick. This distributes spores throughout the area. Wherever they find a suitable place to germinate, new mushrooms will develop. When doing this you should pay attention to the direction of the wind, otherwise you will very quickly find yourself standing in a fine cloud of mushroom spores. You can also hang the bag on a tree branch, where it will be protected from the rain. A spruce with wide, overarching branches, which is growing in an elevated location, is best for this. The wide-reaching branches protect the bag of mushrooms from the rain effectively and the high elevation allows the spores to be well distributed by the wind. If you make a simple device to hit the bag, then there is practically no work left for you to do. A piece of metal to catch the wind, with a strip of wood nailed to it to hit the bag, will perform the job well. The wood will be pushed by the wind and will continue to strike the mushroom bag, which will release spores that will in turn be carried by the wind. With this method I can easily increase the number of mushrooms on my land. It also means that we find mushrooms in the most unlikely places on the Krameterhof.

Many things are possible with wild mushroom cultivation, and there are many areas that have not yet been investigated. Experimentation is, as always, the most important thing. As soon as you begin to work in this area and attempt something new, you will start to understand the causal relationships. In my experience, you will not have to wait long for your first successes.

Substrate of sawdust around birch boletes (*Leccinum scabrum*).

Birch boletes (*Leccinum scabrum*) introduced on an island with birch trees.

DISTRIBUTING SPORES

Wind

Spores

The bag is hung on a
branch where it is sheltered
from the rain.

The slightly elevated
location allows for a
good distribution.

The number of
mushrooms in the area
can be increased
easily with this method.

Spores

In the media a method of harvesting mushrooms is being recommended that, in my opinion, is completely wrong. According to this method, the mushrooms should be cut (the stem is removed), cleaned and the remaining parts of the mushroom should be left behind. On my tours and at my lectures people regularly ask me if this is the best method. In response I must explain some basic facts about harvesting mushrooms: we gather and eat the unripe fruiting bodies of fungi, because the ripe ones are inedible. So, because of this false information from the media, the unripe mushrooms are cut and the rest of the mushroom is left there to rot. The mould spreads very easily in the areas of the fungus damaged by harvesting. Within one to two years the entire mycelium is contaminated and damaged, and the mushrooms disappear.

I observed this effect for the first time in my youth when I grew a variety of cultivated mushrooms such as shaggy ink caps and button mushrooms. It would be much the same if I peeled an apple and threw the peel back in the apple barrel when it was full. The mould would spread and all of the apples would begin to rot. Although almost everyone knows this, many people do not stop to think when they are picking mushrooms and they destroy many good mushroom sites without realising. It would be acceptable simply to pick the mushrooms and clean them at home. If it is absolutely necessary and the mushrooms cannot be gathered so easily, they should at least be cut close to the ground. Then the damaged area should be covered with forest soil, so that the fungus can repair itself. The mycelium will recede and, under good conditions, it will quickly grow new fruiting bodies. As false information about harvesting mushrooms is being spread with such insistence, I get the impression that the intention is to continue to suppress wild mushrooms, so that imported cultivated mushrooms can replace them.

5 Gardens

Kitchen Gardens

The nicest areas around houses were once reserved for kitchen gardens. There, farmers cultivated valuable fruit, vegetables and medicinal and culinary herbs, which meant that they were available right on the doorstep. The purpose of a kitchen garden was not only to grow food; it also served as a pharmacy just outside the house, which was very important for the health of the family. Even as young children we came to see the garden as an important part of our lives. We watched our parents as they worked and we could experience the way the many colourful, sweet-smelling and delicious plants grew.

I can still remember the joy I felt when I pulled up my first baby carrots and radishes in the garden. My mother scolded me, because the vegetables were too small to be harvested, but I simply could not resist. They tasted so good that I pilfered a few from the garden every now and then anyway. As children we were always happy gardening, because there was so much to see: all manner of insects, from earwigs and ladybirds to bumble bees and butterflies could be found there. The garden was full of buzzing creatures flying around, the plants smelt wonderful and we could always find something to eat. We found it so interesting that we always went there in the hope of discovering something new. In hindsight, the most important thing was that we, in a manner of speaking, grew up around the plants and that we could experience the way everything lived and thrived for ourselves. The days were usually too short and it was often dark before we had finished investigating the garden. On these expeditions through our *Gachtl*, which is what gardens are called in Lungau, we also learned how each of the plants were cultivated and where they grew best. We grew up with nature around us and we learnt through play. We could see the way that everything grew, flowered and smelt so wonderfully and also the way it could be prepared into good food. They were gardens for the heart and soul and for the health and well-being of the whole family. These days a kitchen garden like that is usually described as a 'therapeutic garden'. As a result of increasing mechanisation, many farmers turn the areas around their farmhouses into parking places and garages or they build roads around them. By the 50s and 60s this tendency had developed to such an extent that many farmers were even ripping out their old grain silos and existing storage buildings. Old bread baking ovens, which were once built outside, gave way to asphalt parking places. Sadly, many kitchen gardens also disappeared. Very few people were willing to take on any of the work involved in keeping a garden of their own. Fortunately, people

The garden was right by the wall on the east side of the Krameterhof.

are now beginning to think differently. Many people are once more becoming aware of the fact that the quality and flavour of home grown organic produce is far superior to the food bought from a supermarket.

In these fast-paced times when so many people anxiously rush their way through life, more and more will discover gardening to be a relaxing balance to their working lives. For many people their own small garden is their only opportunity to come into direct contact with nature. Happily, medicinal and culinary herbs are also returning to the garden. The healing properties of many medicinal plants have been scientifically proven and are used in modern complementary medicine. This development within the last few years gives me the hope that even more people will soon take an interest in nature and feel a part of it again, instead of believing that they can control it. Creating your own garden is exactly the right way to begin.

Memories of our *Gachtl*

Our *Gachtl* directly bordered the eastern side of the house, where it remains to this day. It was enclosed with a picket fence and had a variety of fruit bushes. I remember the redcurrants, blackcurrants and white currants by the fence and the strawberries that reached all the way to the wall of the house. A gooseberry bush and an intensely sweet-smelling double-flowered rose bush grew in the sunniest part of the garden. This was the best place for them, because these bushes are very susceptible to mildew; the reasoning being that damp, which lasts longer in the shade, encourages mildew. In the dry and stony places we grew thyme, lavender and sage. In the nutrient-rich places we planted mint, lemon balm, sun bonnet, motherwort and lovage, all of which can also cope with light shade. Between these herbs grew poisonous medicinal herbs like monkshood and foxgloves, which catch the eye with their very beautiful flowers. Our mother told us again and again: "You must not touch or eat these herbs, they are poisonous". Poisonous plants are rarely found in gardens today. Possibly because people are afraid that, if left unattended, children might eat them and poison themselves. When I was older, I discovered through my various experiments that poisonous plants play an important role in the interactions within nature. Now I am convinced that they make a significant contribution to a healthy soil life. A varied diet is, in my opinion, incredibly important for the development of soil organisms. After all, an earthworm cannot go to see a vet. The normal and medicinal plants available to animals – however small they might be – should be as diverse as possible. I also think it is very important for children to learn something about the medicinal and poisonous properties of plants, the things that they hear and learn makes an impression on them and influences the way they deal with nature in later life.

To the left and right of the garden gate grew the edible plants that my mother used the most: lovage, chives, leeks, onions and garlic. They were planted here so that they were very quick to reach, as she did not have much time to cook.

Lovage (*Levisticum officinale*) thrives in partial shade and in deep soils. A single plant will cover the needs of a family of four. This popular medicinal and culinary herb hinders the growth of neighbouring plants and spreads vigorously, so it is best to plant it alone in its own corner of the garden.

Naturally, she had to work in the fields and with the livestock too. She would frequently pass the job of quickly going into the garden to collect chives or other herbs to us children – the soup was often already on the table and everyone had arrived to eat.

On the sunny side of the garden were beds for vegetables like beans or peas. If the soil was warm enough, we planted the runner beans in the middle of May because of the altitude (the garden in the Krameterhof is 1,300m above sea level). Mother planted lettuce between them to protect the beans, because they are sensitive to the cold. Lettuce does not present any competition to beans. As a catch crop radishes and carrots are also very suitable. In the other sunny and nutrient-rich beds we grew kohlrabi, cabbages, turnips, radishes and broccoli. Salad plants could always be found as a catch crop: butterhead lettuce, iceberg lettuce, loose-leaf lettuce and endive. However, my mother always kept parsley away from the salad plants: "It doesn't go with the rest," she said.

By the wall of the house stood a damson seedling (*Prunus domestica subsp. insititia*), which we did not prune. The fruit was ungrafted, which means that

Various types of lettuce provide a fresh source of vitamins from spring to winter.

Lemon thyme (*Thymus citriodorus*) develops an intense flavour in dry, stony or sandy areas.

the suckers went on to grow into new trees that produce the same fruit as the parent tree. The quality of this fruit was excellent, it was very aromatic and ripened from the end of September to the beginning of October.

The areas around the fence and in the garden (sunny, shady, dry or wet soils) were given to the plants that were suited to them. This is, without doubt, the recipe for success in any garden, because plants that are in the right places will thrive and are not as susceptible to disease. They also develop the highest nutrient content (essential oils, bitter substances) in the locations that they are naturally suited to. For instance, if you plant thyme appropriately to its natural environment in a warm, dry place (sandy or stony), then naturally it will not grow as high as it would on good garden soil, but it will develop a more intense flavour, which means that its nutrient content will increase. Although thyme cultivated on good garden soil will grow up to 30cm high, it will grow spindly and will have little flavour. The spindly thyme will also not possess the healing properties that many people expect from it. Next to the thyme we grew both sage and lavender. It was certainly an impressive array of scents for such a small area!

The Pharmacy on the Doorstep

The wide selection of medicinal herbs turned kitchen gardens into an indispensable source of valuable medicines for every farm. This was useful, because doctors and midwives were often difficult to reach and also took a long time to arrive. Farmers tended to ask themselves very carefully if they needed a doctor at all, because they could not easily afford this 'luxury'. So in every kitchen garden there was an even mixture of medicinal herbs that might be needed. Every farmer had their own recipes for medicinal creams, tinctures, compresses, poultices and teas. The farmers passed these recipes down through the generations, mainly within the family, and they constantly improved them. This is why remedies vary so much from farm to farm. If people with specific ailments lived on the farm – such as those in need of permanent care – then the farmers would take these particular needs into account when choosing the medicinal herbs.

If someone in our family fell ill, the first thing my mother did, was to go into the garden. For every ailment she knew a herb, which she used in various different ways. She made a tea from mint, lemon balm and marsh mallow and coughs would disappear. Since then the soothing effect of marsh mallow (*Althaea officinalis*) on sore throats, hoarseness and dry coughs has been thoroughly investigated and scientifically recognised.

The herbs were not only used for acute complaints and as medicine, they were also used for cooking. My mother used more or fewer medicinal herbs (lovage, thyme, garlic etc.) according to taste and the health of the family, in a variety of dishes. Many of these herbs are only known as culinary herbs today. These plants, which are mostly used these days without realising, are very important medicinal herbs. Lovage, for instance, encourages the appetite, stimulates digestion and has a diuretic effect. When thyme is freshly cut it has an antibacterial effect and its ability to regulate means that it makes dishes, especially meat and sausage based ones, easier to digest. Perhaps this is why the flavour of thyme complements these dishes so well? Freshly cut garlic has antibacterial and antifungal properties. Eating garlic regularly can even lower cholesterol levels. Finally, it is also an excellent medicinal plant for preventing thrombosis, because it helps to prevent blood clots. The antifungal properties of garlic can also help to protect plants: garlic tea (brew a few cloves of crushed garlic for a short period of time and leave them for a day) can be very effective against all kinds of fungal diseases (e.g. mildew), and lice are discouraged by the pungent smell.

Medicinal herbs were also used to treat sick livestock on farms. For instance, the farmers on practically every farm made their own calendula cream. It alleviates every kind of injury by encouraging wounds to heal and bringing down inflammation. The farmers often successfully used it to treat udder inflammation. Calendula was made into a tea and then used to clean wounds. Since then I have found that calendula also has a beneficial effect on the soil: the plants secrete substances from their roots which discourage nematodes, which (when there are large numbers) can be very harmful to crops plants. For this reason I continue to sow these effective and also beautiful medicinal plants in different areas – preferably on deep, wet soil – and collect the curled seeds in autumn for sowing the following year.

Valerian is another example. Its relaxing properties are well known. Valerian tea was used in veterinary medicine for the treatment of colic and cramps. Cats are an exception to this, because they react very sensitively to valerian. Also chamomile, which is calming and works against cramps and flatulence, does not only help people with digestive problems, but also horses, dogs and chickens.

Many medicinal herbs that were used grew outside the garden on the edges of paths and fields and on slopes. Mugwort, mullein, comfrey, greater celandine,

Purple coneflower (*Echinacea purpurea*) is not only wonderful to look at, but it is also a medicinal plant of considerable value. It strengthens the immune system and is therefore used for colds and to heal wounds.

stinging nettles, lady's mantle, coltsfoot, dandelions, tormentil, cranesbill and chicory are just a few examples.

Because of their inconspicuous appearance they are not seen to be what they are: something special. Now their medicinal properties have been almost forgotten!

Preparations made from medicinal and wild herbs were still widespread in the 40s and 50s. In the following years fast-acting and at first glance effective tablets began to be accepted on even the remote farms and they replaced medicinal herbs. Fortunately – after many people have had to struggle with the side-effects of medication and some have had to take tablets in order to overcome the side-effects of other tablets – we are now beginning to remember this old knowledge, which has been handed down for generations. Unfortunately, over the years many recipes have been irrevocably lost. When I was growing up I got to know many practitioners of natural medicine. When we children had a cough or stomach pains my mother often collected poultices and salves from our grandmother, who lived in Sauerfeld near Tamsweg. A poultice is a mixture made of natural products that is spread on baking paper (or similar) and placed on the chest or back of the person who is ill. The 'montana salve' was particularly effective. It was used to quickly alleviate whooping cough and colds. The farmers made it from the flower petals of different medicinal herbs, of which the peony or 'montana rose' (red flowered) made up the greatest part. This salve had such an intense and pleasant scent that we children never resisted when we were smeared with it or had it applied as a poultice. However, we behaved very differently to another very effective method: applying roasted onion slices, garlic and horseradish. We made this with lard and applied it with hot cloths. The medicinal properties of these methods were astonishing.

Many farmers also made drawing salves. To make this they used tree resin, in other words, liquid larch pitch. They mixed this with different medicinal herbs and made it into a poultice. I remember that the effect of the drawing salve was often so intense that the paper had to be removed, because the sensation became unbearable. The effect was so powerful that it could treat inflammation and festering wounds in a very short time.

Finally, the farmers also made a bone salve from actual bones. For this they saved the bones from cattle and pigs throughout the year in a specially prepared chest. In the chest there was a ventilation grate to allow plenty of air to circulate so the bones could dry out; this grate kept the chest from being invaded by mice. The bones were smoked, because for storage reasons most meat was smoked. In late autumn the bone salve man (*Beinsalbenbrennermandl*, literally 'bone salve burning man') came. This was usually a retired farmer, a woodcutter or herdsman who earned a few schillings as additional income for his autumn years by making bone salve. We children were always happy when this man arrived, because he told us so many stories from his life. We helped to crush the bones so that they would fit into the cast iron pots. Two ten-litre cast iron pots were used. The crushed bones were put in one of these pots and a wire grate was placed on top. In the second pot, which was the same size, we emptied a mug of water (a quarter of a litre). We buried this pot in the wet, mossy soil at a distance from the house. The pot was flush with the soil and upright. We placed the first pot with the bones inside it upside down with the grate facing down, on top of the second one, which was buried in the ground. The grate was only there to keep the bones in. We sealed the space around the two pots with clay and wet earth. Then the *Brennermandl* ('burning man') laid wood over the covered pots and started a fire. Experience was needed to do this, because there could not be too much or too little heat: it had to be exactly right. Naturally, we children wanted to put more wood on the fire to make it as big as possible. If we tried this he would rap us on the finger with a piece of firewood and tell us why we were not allowed. As I have already mentioned, a specific temperature had to be maintained for the fat to be drained from the bones and for it not to be burnt by too much heat. It required great care to ensure that the seal stayed intact and wet. If it were to become damaged, sparks could have reached the steaming oil inside the pots and caused an explosion. At the end of the process, there would be a glutinous brown mass in the bottom pot and in the top pot only light grey flakes of used up bone.

We used this bone salve to treat wounded livestock. The men who came to castrate the pigs, for instance, usually had a salve of this kind. Because of the

MAKING BONE SALVE

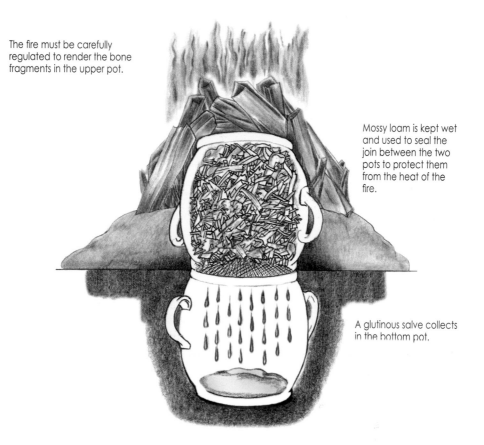

The fire must be carefully regulated to render the bone fragments in the upper pot.

Mossy loam is kept wet and used to seal the join between the two pots to protect them from the heat of the fire.

A glutinous salve collects in the bottom pot.

pungent smell, which is similar to mineral naphtha or wood tar, it was rarely used on people. During the summer we spread a watered down form of it on the draught animals with a rag at haymaking and harvest times, to protect them from flies and horseflies. This provided the animals with very good protection, so that our work could be carried out without interruption.

Through experimentation I have discovered further possible uses for this product, e.g. as a deterrent against bark stripping in forest cultures, or to protect fruit trees from being gnawed by rodents. This remedy provides excellent protection for many years. The bone salve can be mixed with linseed oil, fresh cow dung, slaked lime and very fine quartz sand until it is of a paintable consistency.

You can still make this salve for yourself, but you will need to obtain the requisite bones from a slaughterhouse. These must be placed on a grill and smoked. (We used smoked bones, because we smoked most of the meat to make it last longer. Naturally, we did not have fridges and freezers back then.) Whether the salve would be as effective medically with unsmoked bones, I cannot say. We used the rest of the burnt bones in the garden as fertiliser.

I would now like to describe a few very simple recipes for remedies, which can be also be made by people with small gardens without any great difficulty. There was a time when these remedies could be found in almost every 'home pharmacy'. As the potency of medicinal plants can vary from place to place, the recipes should be adapted. With a little experience the correct strength can easily be determined.

Calendula Salve

When making this salve the whole plant including the stem, leaves and flowers is used. First, two heaped double handfuls of calendula (*Calendula officinalis*) are cut finely. Heat roughly half a litre of lard (available from a butcher) and carefully fry the calendula whilst keeping it moving. Other fats or vegetable oils can also be used (e.g. olive oil). The mixture is covered and left to stand for a day. Then it is lightly warmed, filtered through a cloth and poured into a container. If you make it with vegetable oil, you must first incorporate a thickening agent such as beeswax. For one litre of oil you should use between 200g and 250g wax, which has already been warmed and melted. Mix the melted wax into the filtered calendula oil well, and then leave it to cool. The more wax you use, the stiffer the salve will be; so, if you prefer a very creamy salve you should use less wax. Calendula salve can be used to treat all kinds of injuries, because it encourages wounds to heal and keeps inflammation down.

Sage (*Salvia officinalis*) just before flowering. Its nutrient content is greatest when it grows in a sunny place and without fertiliser. Sage tea is a well-established remedy for a sore mouth or throat and helps with digestion problems.

Lemon Thyme and Thyme Oil

Sprigs of thyme should be picked during dry weather at midday, because that is when the scent is most intense. They should then be placed in a bottle with cold pressed sunflower or olive oil. The oil should be about two fingers above the sprigs of thyme. The bottle is now left in a sunny place such as a windowsill for 14 days. Then the flowers are sieved out with a cloth. To make the oil more potent, the process can be repeated with new flowers. The oil should only be used on children with caution. Keep an eye out for possible reactions on sensitive skin. This old remedy is recommended for sprains and rheumatism; it should be rubbed in regularly to the affected area. It is also recommended for people who have suffered strokes.

Chicory Tea for Diabetics

Steep equal amounts of chicory root (*Cichorium intybus*), dandelion root, stinging nettles, French lilac and bilberry leaves in hot water. For three tablespoons of the plants you will need one litre of water. The tea should only brew for a short period of time and can be taken every day. Chicory was once used by diabetics. The farmers also used to make fresh juice from it, only a teaspoonful of which was taken to lower blood sugar levels.

Tormentil

A powder can be made from dried tormentil roots (*Potentilla erecta*) and stored in a jar. They can, for instance, be ground up in a coffee mill; the finer the powder the better. As a result of its ability to stop bleeding, it is used to treat heavily bleeding wounds. The powder is used directly. The wounds heal very well without leaving large scars.

Vegetable Patch

In addition to a kitchen garden, many farmers also had a large vegetable patch that was fenced off and, like the garden, was replanted each year. In the vegetable patch we grew white cabbage, which was used to make sauerkraut, and provided us with the vitamins we needed in winter. Farmers also planted turnips, chard, beetroot, swede, stock feed carrots and black radish. Turnips, chard and stock feed carrots were used to feed the cattle. We could hardly wait until mother finally put the first *Krautspeck* (bacon cooked and smoked with sauerkraut) on the table. The whole house and surrounding area smelled of freshly cooked *Krautspeck* and sauerkraut. When the postman came to the door he would cry loudly, "Ah, it's *Krautspeck* today!" Naturally, mother could do nothing less than to invite him to have a good portion.

The Most Important Work in Our Gachtl

When I was a child, we always broke up the soil in the garden in spring. This was particularly tiring work for us. Afterwards, we divided up the beds with straight paths. We then positioned the young plants in these beds. The plants had to be planted early to give them a head start in harsh Lungau. They could also go in a container on the windowsill or in a cold frame to start them off.

A cold frame consists of a simple wooden box that is covered with a window or clear sheeting. In the spring, we would place a 30-cm-thick layer of straw and dung at the bottom of the cold frame and cover it with garden soil. The dung warms up through the process of decomposition and functions like underfloor heating for the bed. The cover of glass or sheeting has the same effect as a greenhouse. When creating a bed like this, you should choose a location that is sheltered from the wind and as sunny as possible, in order to make the best use of the spring sun. The plants selected must, of course, be made hardier before they are planted out. Doing this in stages is particularly important so that any damage to the plants' growth is avoided. The plants must get used to the harsh temperatures outside gradually. The easiest way to harden the plants up is by increasing the periods of time the cover is removed. Towards the end of the process, the cover can even be left open by a crack overnight. My mother would begin to harden off the first plants around Saint Joseph's Day (19th March). As soon as the plants were large enough and the overnight frosts were over, she planted them out in the garden. She pushed dry branches into the soil for the peas and beans to climb. She maintained the garden borders, where different bushes, medicinal herbs and flowering shrubs grew. She removed dead flowers and stems, spreading them over the soil around the plants. She then covered the material with a few spades' worth of soil. From time to time, she thinned out the plants by digging up those that were growing too close together and either planting them elsewhere or giving them to the neighbours.

Our garden was very large and the vegetable patch was even larger, which naturally made it a great deal of work. As my mother could not look after the vegetable patch on her own, us children had to help with the raking and weeding. Raking was certainly not our favourite job, although I enjoyed weeding. Sometimes mother would just pull up the larger 'weeds' and leave them amongst the plants – usually on a sunny day so that the roots dried out quickly. In this respect, my small garden was somewhat untidier than my mother would have liked, which made her wonder how everything grew on my dry *Beißwurmboanling* (a steep and stony slope, a *Boanling* is the edge of a meadow). She said that she could save a great deal of work with my method, because the plants would grow just as well or even better, but she could not use it, because the neighbours and her friends would say that the garden was 'untidy'. So we raked and weeded it industriously.

In autumn we harvested the winter vegetables. We pulled them up and put them into piles. Then we took a wooden chair and a chopping block to use as a work surface. We cut off the roots and leaves with a knife. This was done carefully, because the crops must not be damaged, otherwise they would start to rot in the cellar. The storage cellar was a frost-free earth cellar under the house. It was separated into rooms with larch posts. Each crop, such as potatoes, turnips or chard, went in a different room.

In autumn the cabbages went into the large fermentation barrel in the cellar – a large wooden barrel which was sunk into the earth. On the front wall of the cellar was a large pile of sand. We placed the best cabbages from the garden, complete with their roots, in here. We obtained the seeds for the next year from these plants. On special occasions, such as Christmas, we would cook one of the heads of cabbage.

We were very happy when there was a fresh cabbage salad with the Christmas roast (usually pork, prepared in the oven with potatoes and flavoured with garlic, caraway, thyme and marjoram). When we came back from church we could smell the roast some metres away before we reached the house and we would run happily into the kitchen shouting, "We're having roast today!" Having this roast and fresh cabbage salad was very unusual back then; there were no fridges or freezers, and people certainly did not have meat every day.

After the cabbages were removed, the stalks would begin to sprout again They would turn completely yellow from the lack of light in the cellar. As children, although we were strictly forbidden, we always wanted to get hold of them because they were so delicious. Mother needed the roots and stalks for replanting the garden in spring. From the roots and stalks strong shoots would grow, which then grew flowers and seeds. Once the seeds had ripened mother cut the whole plant including the stem, put it in a bag and hung it up in the loft. This way the seeds could ripen and dry out. When the pods opened, the seeds would collect in the bag. All she had to do to sow the plants in spring was to beat the bag a few times against a tree or rock. The rest of the seeds would fall out and then she removed the dry stems. As well as salad and vegetables, the garden provided us with many medicinal herbs, which we used fresh or we dried or pickled them for winter. We also preserved fruit and berries: we dried them, made jam, juice, and schnapps from them or put them in vinegar. Then, as previously mentioned, we harvested and dried the seeds in the garden. We dried the strawflowers for floral arrangements throughout the year, for instance for church festivities. In winter there were few opportunities to get fresh flowers. We were also very careful with our money. So we made full use of the garden; and apart from the tools for working it, we did not have to buy anything. Seeds, young plants, dung and liquid fertiliser were already on the farm, we did not need anything else.

Although I want to preserve and reintroduce old farming techniques, not all the ones that we used were actually necessary. Today I maintain the garden with much less effort. My childlike methods of dealing with unwanted plants have found their place in the main garden. I make sure that none of the soil is left

I control rival plants and keep the soil moist with mulch and full plant cover. This means that I do not have to water the plants or do any weeding. Fresh material should not be applied too thickly. The mulch layer that can be seen here is being spread out.

bare. I achieve this with mulch, by weeding and then leaving the 'weeds' on the soil as well as ensuring full plant cover.

My work in the garden is limited to lightly and carefully working and loosening the soil in the spring and repairing raised beds when required. It is not necessary to dig over the soil to introduce manure, because a nutrient-rich layer of humus develops from the plants that have been pulled up and left. Digging soil over is a particularly bad idea in autumn, because it leaves the loosened soil unprotected against winter frosts. This means that the soil life will not have the necessary protection, so it must either leave or freeze to death. However, I try to protect the soil from the frosts in winter, so I leave the plant cover in place. This protection provides as much warmth for the soil and soil life in winter as a winter coat does for me. Also, the soil does not freeze as quickly, which means that my 'helpers' can work for longer. In nature this works in exactly the same way. In autumn the trees shed their leaves and they collect on the ground like a blanket. Even if the leaves fell for another reason at first, I am convinced that this protective effect of nature is intentional and important. In addition, the biomass remains where it has fallen and turns into valuable humus there, exactly where it is needed.

I think that the familiar method of digging soil over to introduce manure is a bad idea, because cow dung does not work its way 30cm under the ground in nature. Dung always belongs on the surface where there is more air and there are plenty of organisms. Only there can it be properly converted into valuable humus by the soil life. If I introduce manure, I place at most one spade's worth of soil over it, or cover it with a little mulch. There is often too much time invested in working in the garden. So the backaches suffered by those who like digging soil over, should cause them to stop and think. Too much work in the garden does not always bring the success that people hope for and it is not good for their health.

I do not think watering is necessary in the garden either; except during extremely dry weather. With permanent plant cover or mulch, the soil can be protected from drying out. This saves me not only having to water anything, but it also gives me an independent system with independent plants. Excess

Today a fair degree of 'untidiness' prevails in my garden. However, the soil is covered by lush vegetation and is therefore protected from drying out and from the effects of the weather. The soil life is happy and productive.

watering also washes away nutrients, which makes additional manure necessary. We need to escape from this vicious circle and, especially in gardens; we need to free ourselves of this obsession with tidiness, because bare areas of soil are left defenceless against environmental effects.

Natural Fertiliser

Alternative Composting Methods

Composting is a way of producing high-quality fertiliser from organic waste. It is by no means necessary for a high-yield garden to have a compost heap. Mulching throughout the year and the skilful use of polycultures mean that additional organic fertiliser is not necessary. However, anyone who wants to compost anyway, can easily create an unconventional and easy-to-maintain compost heap. To do this, two raised beds running parallel to each other should be positioned so closely to each other that you can only just walk between them. The beds should be built at as steep an angle as possible whilst still holding together (60 to 70 degrees). Organic waste is left between the two beds each day. Each time you do this, cover the waste with a spade's worth of earth, straw, leaves or similar material. Gradually, the organic material will come up to 60 percent of the height of the raised beds. The top layer should be covered with earth and planted or sown with vigorous growing vegetables (pumpkins, cucumbers,

Strong and healthy plants – even without using manure.

turnips etc.). Start at the furthest end of the bed and keep going until the gap is filled up. The best situation is when the dimensions of the compost heap ensure that the gap can be filled and will rot down within a year. The next year you can begin on the opposite side and throw the high-quality compost that was made the previous year on the beds to the left and right with a spade. As many earthworms will be living in there, you should be careful when digging. Afterwards, you can walk through the furrow that is left or climb over one of the raised beds to the side. Planks or large stones can be used to walk on. Using this method you can cultivate vegetables and compost and breed earthworms in a very small space.

Any conceivable type of material can be used for composting: grass clippings, chipped material, leaves, hay, straw, algae or mud from a pond, kitchen waste, cardboard etc. – any organic material that decomposes is suitable. The smaller the material and the more active the soil life is, the faster the compost will become humus. The space between the raised beds is protected from drying out and retains heat, which helps the decomposition process. The plants growing on the raised beds should be chosen so that the compost still gets enough light, but is also protected from the sun. In partial shade the optimal conditions for the decomposition process develop and the compost quickly turns into the highest quality manure.

COMPOSTING BETWEEN RAISED BEDS

Cultivating vegetables, breeding earthworms and composting in a very small space.

Mulch

Mulching also supplies the soil with valuable nutrients. It is nothing other than surface composting; it involves spreading a layer of organic material over the soil to serve as ground cover. The soil receives a protective cover from this, which prevents it from drying out, becoming eroded or suffering from the extreme effects of weather. Leaves, straw, cardboard and plants that have been pulled up whilst weeding are well suited to this. Green manure plants (clover, lupins and mustard) are particularly good. In the mulch layer a constant process of decomposition takes place, through which the mulch is turned into high quality fertiliser. For the material to rot down oxygen is required and the soil also needs it to 'breathe'. When mulching you should also make sure that the material is spread as loosely as possible. If the soil pores become saturated the soil life will suffer.

A loosely spread layer of mulch protects my vegetable plants in the rock garden.

The thickness of the mulch layer depends on the material that I use. I only spread moist or wet material thinly, so that it can rot down slowly and will not begin to moulder. Dry material (such as straw or hay), on the other hand, can be spread much more thickly (20cm or more), because it is looser and has better air circulation. Naturally, it must not be packed down tightly. In addition, dry material does not compact as much as other kinds of biomass, when it rains. In contrast to expert opinion, I do not think that mulch material should be shredded. Experts are probably of this opinion, because the material will rot down more quickly, can work as fertiliser and also is easier to spread around plants. I do not shred the material, because I think it is better that the nutrients are released slowly and that the mulch layer is less likely to compact.

Working with mulch is really very simple: in spring you only need to scoop the mulch a little to the side and you can then sow or plant again. The areas where you sow and plant remain free of rival plants, while the other areas are still protected by mulch. This way unwanted plants can be prevented from growing, while those that have been sown or planted can develop unhindered. With a good mulch cover there is hardly anything left to weed. If you mulch all year round, you must regularly introduce new material. In accordance with the principle of mixed cultures, it is important to vary the plants and materials you use for mulching, otherwise plants will only receive the same nutrients. Variety keeps the soil and plants healthy. As with compost heaps, there are a great number of creatures that live underneath mulch – among them the much loved earthworm. This is one of the reasons that once you have been mulching an area for a while, digging over or loosening the soil in spring will not be necessary. Mulch is also very effective under shrubs, trees and hedges, which is not surprising, because it mirrors what happens in nature already. It was people who first decided that leaves under trees were 'untidy' or 'unattractive'.

Liquid Fertiliser

When I was young, the farmers understood the effect and preparation of liquid fertiliser well. Depending on the effect they wanted and which plants were available, they prepared different mixtures. In this way everyone developed their own 'recipe'. With the appearance of chemical fertilisers and synthetic pesticides, the knowledge of how to use liquid fertiliser has died out in many places. Instead of this many people learn how to spray and fertilise 'correctly' without poisoning themselves. The long-term damage that is caused to our environment by the use of pesticides and chemical fertilisers is not obvious to the majority of people. Unfortunately, many people are willing to accept a short-term increase in yield using these methods. Anyone who wants to treat nature responsibly should say goodbye to the use of chemicals in fields and in gardens. Nature provides plenty of plants that, as a result of the substances they are composed of, are well suited to the production of effective plant feed and liquid fertilisers. To make some plant feed you need to place either freshly cut or dried plants in cold water for one day. Then the feed can be sprayed on your plants. The effect of this can vary greatly. Plant feed made from stinging nettles is particularly popular and can be used universally: the large amounts of nitrogen gives it the effect of good fertiliser and it strengthens the plants. Plant feeds of this kind can be very helpful with vigorous growing vegetables like courgettes, cucumbers and cabbage, however they should not be used with plants with low nutrient requirements like peas and beans, because of the danger of overfertilising. Plant feed made from fresh stinging nettle is also very good against aphids. The aphids seem not to like the smell and the burning effect of the nettle's poison that is retained by the fresh feed is an additional factor. I think that it makes more sense to make cold water plant feed rather than tea, because tea needs to be boiled, which requires a large amount of energy, especially if you want to produce large amounts. I consider boiling unnecessary. If I need stronger plant feed, then I can just leave the plants in the water for longer and stir it regularly. The feed will begin to ferment and turn into liquid fertiliser. Liquid fertilisers are so rich in nutrients that they should always be diluted before they are used. They have – just like cold water plant feed – a good fertilising effect, they strengthen the plants and, therefore, also work naturally to prevent plant diseases, stunted growth and even the prevalence of a single organism. Strong and healthy plants are more resistant to disease; also insects usually prefer weakened plants. These natural plant-based pesticides are very easy to make at home and cost nothing! It is really quite surprising that they have faded into the background so much.

My Method

It is best if locally growing plants are used. It makes no sense to bring plants from a long way away or to import products for this purpose, even if it is recommended in specialist journals. Almost all plants are suitable for making liquid fertiliser. Roots, stems and leaves should only be left in a container long enough for the nutrients to be released, and the liquid fertiliser will develop from these nutrients.

The production of liquid fertiliser for regulating pests must be observed closely over a long period of time. For my plant feed and liquid fertilisers I select plants with that contain certain substances – such as essential oils, bitter substances and poisons. When choosing the plants, I am guided by the instincts and experiences that I have accumulated over the years. Therefore, I continue to try new plants and mixtures, because there is still so much to experiment with and learn about in this area. If I have not used a plant mixture before, I start by making a test tea. For plant feed I use fresh spring water. Tap water is usually artificially processed and sterilised. Also filtration, irradiation, and chlorination can be necessary to comply with drinking water regulations. This water is 'dead' and no longer has any value for me as drinking water. I am, of course, very used to the fresh springs on our farm and I always avoid drinking the water when I come into the vicinity of a town. The taste alone horrifies me. If you have drunk

A well on the doorstep: running spring water is practically a luxury today!

this water for long enough, you probably no longer notice the taste. This works in a similar way to the taste of ripe strawberries and tomatoes that have not been sprayed with pesticides, which people often do not notice any more. If there is no spring water available, you can also use collected rainwater. It is better than treated tap water in any case. You can use any container with a lid; it can be made of wood or even plastic. I do not use metal containers, however, because my plant feed could react with the metal during the fermentation process and produce unwanted by-products. At short intervals (every few days) I test the tea on things like areas of mould, aphids or scale insects and see whether it has the appropriate effect. If the effect is satisfactory the liquid fertiliser is ready and I can use it. If, however, the effect if still too weak I must continue to experiment. So I add more of one plant or another or I leave the tea to ferment for longer. This way more substances will be released and their effect will be intensified. After observing for a long time and experimenting in this way you can create your own recipe for an effective liquid fertiliser, that is the most appropriate to your local conditions.

While the mixture is fermenting it is important that there is enough oxygen in the container. This is why I leave the lid open a little during this time and stir the plant feed regularly with a length of wood. In warmer areas with strong sunlight the fermentation process is much quicker. On the whole, fermentation is complete in a month at the latest in areas that are not particularly sunny. I can tell the liquid fertiliser is ready because it is not foamy any more and it has a dark colour.

I do not think it is necessary to give an exact description of the plant mixture, temperature, quantity of water and plants or the amount to use. The safest and simplest way is to experiment and find out the suitable mixture in the concentration that is necessary for your area for yourself.

For instance, a mixture of plants that I often like to use consists mainly of: nettles (*Urtica dioica, Urtica urens*; they provide nitrogen), and comfrey (*Symphytum officinale* and *Symphytum x uplandicum*; provides potash). I also like to add tansy (*Tanacetum vulgare*), horsetail (*Equisetum arvense*) and wormwood (*Artemisia absinthium*). This liquid fertiliser is effective and improves the hardiness of the plants. It also works against aphid or scale insect infestations and against red spider mites, which is mostly a result of the wormwood. If I have too many of these 'pests' on my plants, I just increase the amount of wormwood until it has the desired effect.

Helpers in the Garden and Regulating Fellow Creatures

I would like to state that, in principle, there is nothing to fight in a healthy environment, because nature is perfect. Therefore, I must think about what effects my system has on nature. If I try to familiarise myself with natural cycles, then a great deal of thoughtlessly carried out actions become unnecessary or even wrong. Every creature has a purpose. The system will only become 'unbalanced' if it is incorrectly managed by human beings. Before you begin to fight 'pests', you should think about the causes of this damaging presence and change the conditions. Problems must be solved at the source. It is not enough just to treat the symptoms.

Here is an example: if I have too many aphids on my fruit trees, that means there are not enough natural predators (among these are ladybirds, earwigs, hoverflies, lacewings, various spiders, beetles and birds) and frequently there is not enough shelter or suitable habitat for them. If, on the other hand, I have a good habitat underneath the trees that are infested with aphids, and the ground is richly structured with stones, branches and leaves, the number of creatures which prey on aphids will increase. They will find an 'open buffet' and the overpopulation of aphids is quickly cut back. It is not necessary to take additional measures.

There was hardly ever an overpopulation of 'pests' in our garden. This was mostly a result of the diversity and good structuring of old kitchen gardens. The more diverse a system is, the more stable it will be. Monocultures are favourable environments for the sudden large-scale appearance of a single kind of creature, because they find a surplus of food. The 'pests' jump from one feed plant to

A crab spider (*Thomisidae*) on a daisy (well camouflaged) first lies in wait for its prey, and then consumes it. In a working food cycle there are no useful or harmful organisms, just fellow creatures – some, such as in this picture, are particularly beautiful.

another, so to speak, because their natural predators do not find the right conditions in this wasteland. In a polyculture these problems can never arise, because there is always a wide variety of plants available. The spread of disease is also checked by this diversity. The valuable, helpful and beneficial creatures need suitable environments and places to hide and hibernate of their own.

These factors ensured that the *Gachtl* of my childhood was protected from large-scale damage resulting from pests. I can only remember a few times when the cabbage white population was much greater and descended on our vegetable patch. This overpopulation can be explained by continuous natural fluctuations in the population of pests and useful creatures. Nature works using the system of supply and demand. An increase in useful creatures balances out the increase in pests again after a while. If you resort to using chemicals in these situations, it will have the opposite effect, because many pests are more resistant to pesticides than their natural predators. So some of the pests will survive the attack and all the useful creatures will die, which can make the following wave of damage far greater. We checked the overpopulation of cabbage whites on the vegetable patch quite simply by spraying a liquid fertiliser composed of wormwood, stinging nettles, gentian root and horsetail on the cabbages.

Some of the most important creatures in the garden are: slow worms, lizards, hedgehogs, birds, amphibians, spiders and predatory mites. There are also many insects such as ladybirds, ground beetles, hoverflies, lacewings, earwigs, ichneumon flies and dragonflies. Only a small amount of energy is required to provide all of these helpers with a suitable habitat. The most important thing is for the garden to be richly structured and to resist making everything straight lines and tidily swept. The creatures need places to hide, nest and hibernate and a wide range of food to be happy. This is exactly what you need to provide them with. The edges of the garden are particularly well suited to this. Here you can, for instance, grow wild fruit and flowering hedges or even a variety of different

The sand lizard (*Lacerta agilis*) likes sunny places, e.g. piles of wood or stones on unsurfaced ground. Thick vegetation in the immediate vicinity (flowering meadow, hedge) is preferable. Its diet consists of insects, spiders, woodlice and slugs, amongst other things.

wild flowers. It is a particularly good idea to put in tree stumps or gnarled hollow tree trunks. These can provide good areas for these creatures to breed and they are also very attractive to look at. Piles of wood, branches or brushwood can also fulfil this purpose.

Birds and bats can be encouraged with nesting boxes and the berries and fruit growing in a wild fruit hedge. Stones or piles of stones can also offer varied habitats, which can even be combined with a herb spiral if used carefully. Areas of water and wetlands enrich a garden greatly, because a population of amphibians and dragonflies can develop there.

Only a small amount of work is required to create sheltered areas of this kind and, with a little creativity, they can make the garden even more pleasant to look at.

Voles

Voles rarely appear in large enough numbers in our garden to cause damage. The reason for this is the following: in the confusion and diversity of plants the voles find enough to eat. They chew the roots of many different plants and shrubs; there is, however, no complete crop failure, because there is enough food for everyone. In the areas that have been gnawed on, the individual shrubs can repair themselves quickly and many new fibrous roots will grow around them.

The voles also take many pieces of root away and store them for the winter, or feed them to their young. However, they regularly lose individual pieces of root in their extensive networks of tunnels. These tunnels are collapsed by rain or are colonised by other animals, and the voles must rebuild them. The lost salsify, black salsify, Jerusalem artichoke and carrot roots, to name a few, begin to sprout in the tunnels and new plants develop in the most unlikely places and in the most inhospitable areas. These are frequently places where you would never have thought to plant anything. The tunnels themselves drain off excess water and aerate the soil.

Many insects, plants and animals are territorial: they claim certain areas as their territory and defend them. According to my observations and experience it does not make any sense to fight voles, because once the territory becomes empty it will be used by new voles coming to the area. If I fight them (with poison, gas or by catching them), the territory will only become free for others. The lower population density will be balanced out by more and more empty territories. Voles will produce more offspring or even just produce more males. Instead of catching, poisoning or gassing pests, it is better to consider the cycles of nature. If I let the voles work for me, I will have aerated, loose and well-drained soil and also lush, diverse vegetation. The vole will no longer appear as a cause of damage. Moreover, poisoning and gassing will contaminate the soil. If the voles are exterminated on a large scale, the soil will no longer be well-drained or aerated; it will harden and become more acid and mossy. This will lead to many plants losing their habitat. The energy required to repair damage to the soil is much greater than the supposed damage caused by the voles eating crops. It is important to make sure that there are always enough decoy plants available to them. Decoy plants are particularly tasty plants, which the animals prefer to eat. Jerusalem artichoke and black salsify make very good decoy plants. If there are enough available, the voles will leave the fruit trees alone. It is not a question of what can I do to fight the 'pests', but what can I do for them, so that they will not cause damage and even work to my benefit.

Slugs and Snails

The situation is different with the non-indigenous Spanish slug (*Arion vulgaris*). Where we live the slugs breed on an enormous scale; in many places people have little idea of how to deal with this menace. Whilst giving consultations in Southern Styria and Lower Austria I discovered that on farms and in places where vegetables were being grown there were up to 15 slugs per square metre. Many farmers complained that the cattle would no longer graze, because the grass was so full of them. "Growing vegetables without using slug pellets is no longer an option," was the opinion of the troubled landowners. The owners of gardens in town told me that the slugs would crawl up the houses all the way up to the balconies. In many cases, espalier trees and climbing plants had to be removed from house walls to discourage this.

A lush growth of decoy plants (here mostly Jerusalem artichokes) protect a newly planted orchard. Also to be seen are foxgloves (*Digitalis purpurea*), a highly poisonous medicinal plant (not to be used for self-medication), which I plant to improve the health of the soil among other reasons.

In smaller gardens the following method is very effective in my experience: take a watering can, cut the spout to half its original length so that it is much wider. Fill the watering can with a mixture of very dry fine sawdust, ideally collected from a carpenter or joiner's workshop. The sawdust must, of course, come from untreated natural wood and not be varnished or contain any other harmful substances. I take the sawdust from a carpenter's workshop, because the wood there is completely dry and the sawdust is much finer than you would find in a sawmill. Moreover, sawmills mostly work with fresh wood. I mix the sawdust with one part wood ash to ten parts sawdust, or with quicklime powder (around 1:20). Alternatively, you could use both, the only important thing is that all of the ingredients are bone dry. I fill the watering can with these materials and pour a finger's width border of the mixture around the outside of the lettuce or vegetable patch. Make sure to free the border area of vegetation first. This border of sawdust mixture should remain as dry as possible. This means that from time to time, especially after is has rained, you will have to replace it. The fine dry sawdust mixture adheres to the foot of a slug or snail the moment it tries to get to the lettuce or vegetable patch. The ash and quicklime extract moisture, which prevents them from getting into the crop. If you sit in the garden in the evening, you will be able to see how the slugs and snails turn around when they reach this barrier and go back the way they came. Successes like these will quickly take the fear out of a slug or snail invasion.

There are many ways to regulate these pests naturally. Here is one more. Slugs and snails lay their eggs in dark, moist places. If you provide them with an ideal habitat to lay their eggs, you can regulate their population. To do this I make rows of freshly cut grass and leaves in the garden. They should be piled higher and compacted more than mulch, and should be kept as moist as possible, so that they provide the best conditions for egg laying. Slugs and snails will travel great distances to use places like these. On a particularly sunny day I then go into the garden and turn over the rows of grass with a gardening fork. Whole clusters of eggs will have adhered to the rotting grass. If you turn the rows of grass over at midday when it is at its sunniest, the eggs will rapidly be destroyed by the heat of the sun and the UV rays. The overpopulation of slugs and snails can quickly be counteracted with this method. If your neighbours also use it, the effect will be even greater. This method also demonstrates how much damage the improper use of mulch (using fresh material, piling it too high and not loosely enough) can cause.

Apart from these measures it is, as already mentioned, important to have the natural predators of slugs and snails as helpers in the garden. Excellent examples of these are hedgehogs, shrews, lizards, toads and many kinds of ground beetle.

The well-known edible snail (*Helix pomatia*) also helps to regulate the frequently large populations of slugs by eating their eggs. So not all snails are harmful!

Earthworms – Nature's Ploughs

Earthworms are among the most important helpers in every garden. We have many local varieties on the Krameterhof. The brandling worm (*Eisenia foetida*), common earthworm (*Lumbricus terrestris*) and red earthworm (*Lumbricus rubellus*) tend to appear in large numbers in healthy soils. You can easily recognise the brandling worm by its dark red colour and its distinctive yellow bands. Common and red earthworms do not have this distinctive banding. Brandling worms are epigeal, i.e. they live on the soil surface. Red earthworms, on the other hand, only spend their youth on the surface and later they burrow into the deeper soil layers. Finally, common earthworms, which many think of as 'typical' earthworms, create burrows to live in and search for food at depths of up to three metres.

These three kinds of worm complement each other wonderfully in their work for gardeners: the brandling worm processes large amounts of organic material and provides the best compost. The common and red earthworms reach deeper layers of soil with their tunnels and aerate the soil well. The tunnels also work like an ingenious drainage system. The soil can retain more moisture; it does not dry out as quickly and is better protected from surface erosion. Plant roots can extend through earthworm tunnels better. Naturally, both of these types of worm also produce nutrient-rich compost for the garden. Earthworm casts contain much more of the plant nutrients nitrogen, potassium, phosphorus and calcium than can be found in the best garden soil. With their crumbly consistency they also provide a good soil structure. As a result of these factors, the vegetation develops much better with the help of earthworms. The plants are healthy and therefore more resistant to disease.

This is why it is important to provide the best possible living conditions for these valuable helpers. As earthworms are sensitive to UV rays, it is a good idea to make sure that the garden has permanent soil cover. This can be achieved with a mixed crop that is selected so as to avoid large areas being harvested all at once. Mulch also provides soil cover and attracts earthworms. If you find very few earthworms in your garden, you should by all means attempt to breed them yourself, particularly since this is easily acheivable in very small areas. Breeding earthworms is inexpensive and requires very little time. You can also 'dispose' of your organic waste. As an end product you will receive high quality compost for flower pots and the garden, and many energetic helpers. If you begin breeding earthworms on a large scale, you can even develop an additional source of income through the sale of worm compost and worms. In Europe and the United States there are a number of companies entirely dedicated to the breeding of earthworms.

Breeding Earthworms

In order to breed earthworms successfully you must research their natural habitat. Your own system will be designed accordingly. On a small scale, a wooden box with a capacity of one cubic metre is enough. Earthworms require a soil substrate of a mixture of straw, cardboard, soil and a little dung. In my other attempts I have also used different materials like natural fabrics (cotton, hemp etc.). The soil should be loose and well aerated. To ensure this, it is a good idea to incorporate layers of branches, leaves and roots into the substrate. Any cooking waste can be used as food for the worms. Onions and garlic are the only things I do not give to my earthworms, because I get the impression that they do not like it very much. The worms particularly like used coffee filters complete with coffee grounds. It is important to provide a regular supply of organic matter, so that the worms continue to get fresh food. The amount of food should be adjusted to the amount of earthworms. If the worms can break down their food as quickly as new food accumulates, the rate is optimal and harmful build-ups of mould will be prevented. Room temperature is ideal for the worms. A steady balance of moisture and a good supply of oxygen are also necessary. To avoid a build-up of water, holes should be drilled in the bottom of the worm box. The soil should be neither completely dry or completely wet; too much water will make the worms pale.

You should observe the worms regularly. You will quickly recognise whether or not they are comfortable in their environment. Intuition is an important factor in creating the optimal conditions. In my greenhouses I no longer breed earthworms in boxes, but directly in the soil. To do this I use the soil substrate

Plenty of worms can always be found in good garden soil.

already mentioned, cover it with earth and put some worms in the pile. In the middle of this pile of earth I make a shallow depression. I can put fresh organic waste there each day and cover it with a few handfuls of soil. If when I'm feeding the worms the pile seems too dry, the waste quickly makes it moist again. If the system is designed so that it is well ventilated, feeding the worms every second or third day will be enough; this means that they can be left to their own devices over the weekend quite happily.

Along with the valuable humus and the numerous earthworms and worm eggs, breeding these useful creatures has yet another advantage: you will learn to observe and put yourself in the shoes of other creatures. Your ecological understanding and empathy will improve. From time to time when the weather is wet, I place the worms I have bred along with some soil and worm eggs in a bucket and scatter them over new terraces and raised beds in the evening. I use the nutrient-rich and fine crumbly worm humus for especially valuable and demanding plants and also for the flowers on my balcony.

Characteristics of Town Gardens

How Children Experience Nature

In principle, a garden in town has the same purpose as a kitchen garden. It is my opinion that town gardens are more important today than ever. If you live in a town and do not have the opportunity to grow up around animals and plants in forests and fields, you can at least experience a little nature in your own garden. The size of the garden is of little importance. The therapeutic effect of experiencing the marvel of creation in your own garden is a much more important factor.

I think back to my childhood when I planted my first horse chestnut, which I used to play conkers with, in a window box. My mother said to me, "If you plant that horse chestnut in the soil a tree will grow out of it." She preferred her plants to be in window boxes on the sill in the kitchen instead of in the garden. My horse chestnut developed into a splendid little tree. I cannot describe the effect this had on me, because all of my later successes came back to that one experience. If children have the chance to grow up around nature, then they will be able to learn from it. It is incredible how much there is to discover. Intensive observation will inspire them with ideas that they will want to implement straight away. Learning begins and success will follow. Children do not give up easily, they are curious and have special access to nature. Their urge to discover motivates them to try again and again if they do not succeed the first time – that is the most important thing: to never give up and to learn from your mistakes. Children need praise and to have successes, this makes them strong and encourages creative and independent thought. Children still have room in their heads to retain their own observations and experiences of natural cycles. These memories will stay with them for their whole lives. My childhood experiences have helped me always to come back from the wrong path and find a natural life in harmony with nature. If you isolate children from nature, cut them off from their roots in a manner of speaking, they will not understand causal relationships and cycles within nature. As they have no roots, they will find it harder to handle problems. That is why if you live in town you should still give your child the chance to sow radishes or carrots in the garden or in a window box, and to watch them flourish. This will allow them to make observations about insects and experience the colour and smell of plants. The desire to discover nature exists in every child, if the parents do not educate it out of them or forbid them to go any further into its secrets. How often have I heard:

Children should have the chance to grow up around nature. These are my grandchildren: Helmut, Elias and Alina.

"Don't get dirty, the ground is filthy", or "Come away, leave that alone." The same goes for when children point out a butterfly, a bumblebee or a beetle to their parents. It is not uncommon to hear: "Leave it alone, yuck, that is horrible, come away. It's poisonous, it'll bite you and, anyway, you'll get dirty." In my opinion, this is one of the biggest mistakes a parent can make. You should give the child your attention and ask: "Oh, what have you found there?". Try to find out what kind of worm, beetle or butterfly it is. You could look through a book on insects with your child in the evening and work out what they found. This way your children could grow up having a relationship with nature even if you live in town.

Design Characteristics

On the whole, everything that applies to a kitchen garden can also be applied to a town garden. If there is only a small area available, then it is even more important to design and use it optimally. In small town gardens, for example, valuable space can be gained by creating raised beds and terraces. The same principles apply in this situation as those already described in the chapter 'Landscape Design'. These techniques will provide microclimates, visual barriers, windbreaks and protection from erosion. As a result, incoming pollution will be minimised (especially fine dust) and noise pollution will be mitigated. All of these factors are beneficial for a town garden and should not be underestimated.

Before the landscaping begins, the existing soil must be examined. When doing this all of the factors previously mentioned in the section 'Soil Conditions' are important. It is possible for the soil in town to be so heavily polluted, that you will have to replace it with uncontaminated soil from an organic farm before you can begin work on the garden. Although this is costly, with some soils it is sadly necessary. Over time, an active soil life should develop in this soil, which will find the best conditions and will be encouraged by the use of mixed crops, and the lack of chemical pesticides or chemical fertilisers. The regenerative power of the soil will be enormously improved by this, to the extent that you will be able to grow high-quality produce in town. If the soil is heavy loam, which is water and air permeable, it is possible to loosen and aerate it by mixing in sand, straw, leaves and chipped material (wood chip). If you are using an excavator for landscaping, you must first find out whether there are any telephone cables or gas, water or sewage pipes in the ground and exactly where they are.

When shaping a small garden it is particularly important to make the most of the sunlight. If you do not select your plants carefully, the whole garden will quickly become shaded. This is why tall-growing trees should not be planted. If there is a house or shed wall available, the masonry stove effect of the bricks that I have already described will be at your disposal. The heat retention and radiation qualities of the wall makes it suitable for fruit trees that need a large amount of heat (peach and apricot), and can be planted as espalier trees. A system of tiered terraces – in other words, using vertical surfaces in every possible way – is of

Fruit tree as a 'climbing aid' for tomatoes.

Lush and diverse plant life can even develop on the west side of the Krameterhof, which is in shade.

great advantage in small areas. On the different terraces, shrubs and fruit trees can be planted at staggered heights. The trees can then be used by grapes, kiwi fruit, cucumbers, pumpkins, courgettes, peas and beans as climbing aids. This way the heat retention and radiation aspects of the walls will be used effectively. The interaction between the nutrients released by the individual plants in symbiotic communities of this kind is shown to best advantage. You can create a real 'jungle garden' that offers a place to recuperate and relax, in addition to providing delicious produce. Naturally, you must find out how high the different shrubs and trees will eventually grow before you plant them. This way you will save yourself the work of constantly having to prune and trim everything back.

In gardens where the sunlight reaches areas abruptly because of tower blocks or other buildings, you must make sure that it does not hit any frost-sensitive trees, which are in full flower, too suddenly (e.g. apricot, peach or early

PERMACULTURE IN A TOWN GARDEN

Through skilled use of the space, fruit, vegetables, herbs and mushrooms for personal consumption can also be cultivated in a small area.

cherries). Although these trees can withstand light overnight frosts without taking damage, abrupt sunlight can put them into shock, which can lead to the loss of all their leaves and flowers. In this situation you should place trees in areas where a shock of this kind can be avoided, instead of positioning them by the sunny house wall, which would otherwise be optimal. Overnight frost can thaw slowly in the shade, which has less serious consequences for the tree. The fruit will ripen a little later and might not be as sweet, but this compromise is necessary for there to be a harvest at all.

The conditions that can be found in town gardens vary greatly. This is why it is important always to remember the principles of permaculture and to treat your own patch of land with empathy and creativity. Then you will find plenty of ways to grow vegetables, medicinal and culinary herbs, berries, fruit and mushrooms in only a few square metres.

Terraces and Balcony Gardens

Permaculture principles can be made a priority and put into practice on balconies, terraces, small green areas and even in houses. In fact, I even had a small plant tub in my first 'garden'. I was sceptical at first, but you really can grow anything, no matter how big or small, in a container like that. I have planted up balconies and terraces in many different towns. To start with there are usually just ornamental trees or bushes like cotoneasters, junipers or dwarf Alberta spruces on the terraces or balconies, mainly because they do not need much care and are 'green'. Usually, everything is very homogeneous; this is probably because it is stipulated by the house rules or there is too little flexibility. Balcony, terrace and even normal gardens can be found with hardly any variation in design or plant selection throughout all of Europe. I continue to hear from the owners of gardens like this, that nothing else would be able to grow on the 10th or 20th floor anyway, and certainly not fruit or vegetables in any case! Then they often comment that they do not know what the neighbours would say if they suddenly saw radishes, peas or even beans growing in a plant tub. I encourage people to just break this taboo and make their own balcony or terrace garden into an edible garden regardless. My methods and suggestions have been put into practice successfully again and again.

Let us use the example of a small terrace, two metres by three, that is facing away from the street. At this point I should mention that you need to keep an eye on the amount of pollution coming from roads or factories when cultivating food in town. On busy streets it is better not to use the sides of the house that face the road to produce food. It is also a good idea use the walls of the house by planting climbing plants like clematis. This can also create a microclimate by adding an insulating layer that can cool the house in summer and help retain heat in winter. Areas that are a little more sheltered and at the back of the house are, however, very well suited to the production of food. At the front of the terrace you could place two concrete troughs with a combined

Medicinal and culinary herbs and even vegetables can be cultivated on a small balcony.

capacity of around one and a half cubic metres of soil. Drill one or two holes with a diameter of approximately 10cm in the bottom of each trough. Put some bricks or wooden posts under each trough so that there is a space of around 15 to 20cm between them and the ground. In this space put a waterproof tray. Now you can insert a hardwood trunk in through the hole in the tub. Make sure that the trunk is narrow enough to fit through the hole, whilst still leaving space for water to trickle through. As long as the trunk fits in the space available, it can be as tall as you like. This acts as a climbing aid for grapes, kiwi fruit, courgettes, cucumbers, pumpkins, beans, peas, roses and various other climbing plants and it can also be used for the cultivation of culinary mushrooms, as I have described in the 'Mushroom Cultivation' chapter. If you select a particularly attractive trunk (with pleasingly twisting side branches), the garden will look even more pleasant. Directly around the hole in the trough (around the trunk) place enough broken bricks or gravel to provide drainage and prevent a build-up of water in the trough.

The trunk in the trough can now be drilled in a number of places and inoculated with mushroom mycelium. Then the trough is filled up to around two thirds with healthy soil mixed with broken bricks. You should not use commercial potting soil for this, because it contains large amounts of peat, which is harvested at the cost of our moors and does not have any kind of fertilising effect! Earthworms are also introduced into the trough. Then the planting and sowing can begin. The climbing plants are wrapped around the trunk and various vegetables (lettuce, radishes, peas etc.) can be planted or sown next to them. The more you manage to make use of different levels, the more green material you will be able to fit in a small space. An arrangement of plants at staggered heights achieves this best. Plants that grow to different heights can be positioned so that no competition will arise.

Fill the tray with water. The hardwood trunk (the criteria for selection can be found in the chapter 'Mushroom Cultivation') 'sucks' the water up from the

Colourful combination of plants by the wall of the house.

tray and balances out the moisture of the soil in the trough. If the trough is left outside in the elements, enough rainwater will collect in the tray to keep the trough moist. Otherwise, it will need to be filled up by hand, or the plants will have to be watered. If you have a gutter, you can keep the plants supplied with water automatically. Fix a section of drain pipe into the tray from the gutter and put in an overflow that leads back again (include a sieve, and position the overflow at least 10cm higher than the inlet pipe). However, in cities you should be careful about using this method of irrigation, because the roofs there are often very dirty and ash, soot and many harmful substances could accumulate in the gutters. If, on the other hand, your house is in a less heavily populated area, you can use this method of irrigation and go on holiday without having to worry that your balcony garden will dry out.

Organic waste from the kitchen can also be incorporated into the soil in the troughs each day with a trowel. The waste should always be used fresh and placed in a different area each day. It should be covered with leaves or mulch whilst making sure that plenty of air can reach it. The organic waste provides the worms with food and the plants with high-quality fertiliser. Over time the trough will, of course, fill up and the result will be a substrate with an enormous amount of worm eggs and young worms that can then be used in plant tubs, flowerpots or in the garden.

The liquid fertiliser that I have already described can also be used to protect the plants on a terrace or balcony and increase their resilience (against aphids and fungal diseases like mildew among other things). The mixture you use will

BALCONY GARDEN

A richly structured balcony used in many different ways
makes it possible to experience nature in a town.

depend on the number of plants and the available space. The plants required for the liquid fertiliser can be obtained on a walk through the countryside or through a forest. Some liquid fertilisers develop a very intense odour. If the smell bothers you, something can be done about this easily: simply stir in some stonemeal and the odour will be neutralised, valerian can also be used. If you do not want to make liquid fertiliser you can use herbal tea instead. Chamomile tea, for instance, has an antibacterial effect and can be used to prevent root diseases. Tansy is very effective against root lice and can be used to treat rust. The tea can be used as soon as it has cooled. It is a matter of preference which method you decide to use, because the plant mixtures are just as potent when they are prepared as a tea, an extract or as diluted liquid fertiliser. Your own experiments will, over time, lead you to the best mixture for your balcony.

With time the climbing plants will stabilise and become woody (grapes, kiwi), so that they will no longer need additional support. This means that it is not a problem when the trunks inoculated with mushrooms lose their load-bearing capacity over time.

In exposed areas the plants should be protected over winter. On a balcony the fluctuation between cold and warm temperatures is often particularly extreme. You could, for example, protect the concrete troughs in winter with some jute bags. Sensitive climbing plants should be protected from the overly intense winter sun with the help of rush mats, because sunlight causes most of the frost damage! In addition, the soil in the troughs should be covered with leaves or jute bags over winter to stop it freezing through. This way the earthworms will also survive the colder months.

The yield of mushrooms, grapes, kiwifruit, fruit and vegetables in this small area clearly shows that the plants are constantly supplied with additional nutrients by the intensive work of the earthworms. The 'masonry stove effect' created by the radiation of heat from the house wall also has a very positive

Kiwi plants grow best in sheltered places, especially by the walls of houses: the picture shows a hardy kiwi (*Actinidia arguta*, small fruit) that has already grown up and over the roof of the Krameterhof. The more sensitive, large-fruited kiwi (*Actinidia deliciosa*) can also be cultivated easily.

effect on the growth of the plants. The plants can spread even further with the help of climbing aids, so that a wonderful pergola of leaves develops on the terrace, which can provide shade and serve as a visual barrier. The size of the system can be increased according to preference, there are no limits to your imagination! A further positive effect is that, even if it is on the 20th floor, this garden will give children the opportunity to experience and grow up around a piece of nature. Various butterflies, bumble bees and bees will quickly arrive in these mini gardens. Birds can also make their nests there. The fact that a mini garden like this can also produce a pleasant climate indoors and fill the air with beguiling scents is just incidental. Therefore, urban permaculture does not just represent a nutritious use of space, it also increases the standard of living and promotes garden areas as recreational spaces.

Bypass Technique

Neighbours usually appreciate the beauty of a flourishing permaculture system. If your neighbours also become interested in permaculture, you could create a terrace system reaching from storey to storey. Climbers like grapes and kiwi plants can grow up the front of the building from one flat to the next using a balcony as a climbing aid. On the balconies, troughs with soil can be prepared so that the plants can put down new roots (lead the plant into the trough, heap soil over it and possibly weigh it down with a stone). The plants will then draw fresh strength and nutrients from the soil and will grow from one storey to the next (in other words, the 'bypass technique'). Each storey can be overseen and harvested by the people living there. In this way a communal garden will begin to emerge. If one of the residents goes on holiday or is going to be away for a longer period of time, there is no danger of the system not being able to function. The plants are rooted in containers on different storeys, where they are provided with water and nutrients. A system of this kind is vertical as well as horizontal, as it can travel in any direction. Creative thinking is, of course, needed here, because there are so many possible ways to use, design and plant a system like this. Naturally, attention must be given to the structural integrity of the building with the additional weight of the planters.

With the increasing access to nature, people's empathy for each other will grow. So urban permaculture can improve the climate of the town literally as well as figuratively by promoting interpersonal relationships. Plants can be a bridge between people.

If politicians and businesses put the philosophy of sustainability into practice, then towns and cities could turn into green oases. All courtyards, car parks, playgrounds, open spaces, walls and roofs could gleam with luxuriant green vegetation and people could benefit from positive side effects such as less contamination by dust and harmful substances. To do this we need to change the way we think in a broader sense. Permaculture systems do not function in isolation from outside influences, they demand cooperation!

BYPASS TECHNIQUE

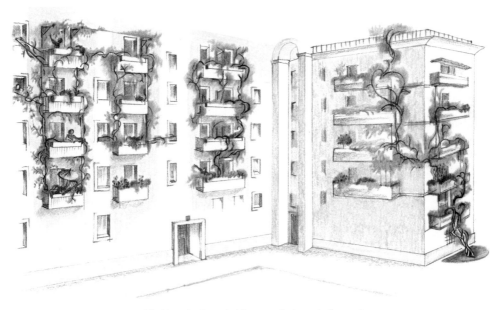

Climbing plants are led from one balcony to the next.
The plants will slowly spread over the entire block of flats.

On each balcony the plant
(here a kiwi plant) will put
down new roots and provide
itself with water and nutrients.

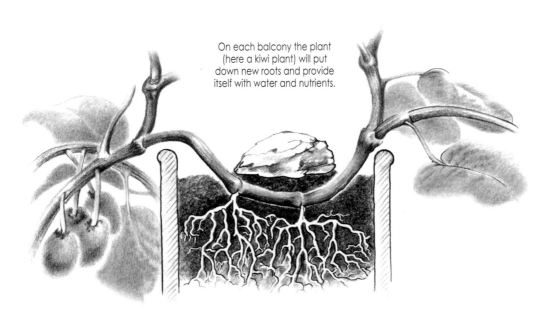

Plant List

The following list will give you an overview of the locations required by various plants and information on favourable plant communities. It is still important to observe the plants in order to find the best combinations for the conditions on your plot of land.

Vegetables

Plant	Supporting Plants	Characteristics and Requirements
Bean (*Phaseolus sp.*)	Savory (a culinary plant with an intense flavour and light effect against aphids), maize (serves as a climbing aid) and many more	Undemanding, belongs to the legumes (increases the amount of nitrogen in the soil), sunny to partial shade, prefers loose soil, low nutrient requirements
Pea (*Pisum sativum*)	Maize (serves as a climbing aid) and many others	Undemanding, belongs to the legumes (increases the amount of nitrogen in the soil), sunny to partial shade, prefers loose soil, low nutrient requirements
Cucumber (*Cucumis sativus*)	Peas, beans, garlic and basil (prevents mildew), good King Henry and Jerusalem artichoke around the edges (windbreak)	Prefers good soil, sheltered areas (sun traps), high nutrient requirements
Good King Henry (*Chenopodium bonus-henricus*)	Good around the edges of pumpkins and cucumbers; serves as a windbreak	Undemanding and hardy, but prefers good soils, sunny to partial shade, frost hardy, wild vegetable
Carrot (*Daucus carota*)	Onions, leeks, peas, beans, lettuce and black savory	Prefers good, loose soils; sunny to partial shade, medium nutrient requirements
Potato (*Solanum tuberosum*)	Marigolds (to discourage nematodes), peas, beans and onions	Prefers good soil (not too wet), sunny to partial shade, high nutrient requirements
Garlic (*Allium sativum*)	Included to discourage fungal diseases in susceptible plants (cucumber); is not particularly choosy itself	Prefers loose, light soils; sunny areas, medium nutrient requirements
Chinese artichoke (*Stachys sieboldii*)	Peas, beans, beetroot and black savory	Prefers good soil, sunny to partial shade, frost hardy, tuber
Cabbage (*Brassica oleracea*)	Peas, beans (to improve the soil), marigolds (to discourage nematodes), southernwood, basil and mint (repels pests with its scent and essential oils), lettuce (ground cover)	Good, moist soil, high nutrient requirements
Pumpkin, courgette (*Cucurbita ssp.*)	Beans, peas (to improve the soil), maize, tomatoes, good King Henry and Jerusalem artichokes around the edges (windbreak)	Prefer good, moist soils, pervasive, sunny areas, high nutrient requirements

Plant	Supporting Plants	Characteristics and Requirements
Leek (*Allium ampeloprasum*)	Carrots, garlic, tomatoes, radishes (to keep pests away), parsnips and skirret	Prefers good, moist soils; high nutrient requirements
Maize (*Zea mays*)	Beans, peas, tomatoes and lettuce (ground cover)	Prefers nutrient-rich, moist soils; sunny areas, high nutrient requirements
Chard (*Beta vulgaris*)	Beans, peas, cabbage, radishes, lettuce and mint	Good, moist soil; good ground cover is particularly beneficial
Pepper (*Capsicum ssp.*)	Tomatoes, leek, lettuce and cucumbers	Good, moist soil; sunny and sheltered areas (sun trap)
Parsnip (*Pastinaca sativa*)	Lettuce, black salsify, onions and leeks	Good, loose soil, frost hardy, tuber
Beetroot (*Beta vulgaris*)	Beans, peas, onions, radishes, lettuce, borage and cabbage	Undemanding, but prefers moist soil, good ground cover is a large advantage; medium nutrient requirements
Lettuce (*Lactuca sativa*)	Radishes, cabbage, kohlrabi, onions, leeks, borage, beans, mint, spinach and many more	No particular needs, medium nutrient requirements
Wild rocket (*Diplotaxis tenuifolia*)	Well suited to being a catch crop or plant cover	Undemanding, sunny to partial shade, annual, vegetable (salad)
Black salsify (*Scorzonera hispanica*)	Onions, garlic, lettuce, carrots; good as a decoy plant near fruit trees for voles	Prefers good, loose soil; low nutrient requirements
Celery (*Apium graveolens*)	Cabbage, peas, beans, leeks and cucumbers	Prefers good, moist soil; high nutrient requirements
Spinach (*Spinacia oleracea*)	Beans, peas, radishes, lettuce and cucumbers	Prefers good, moist soil; medium nutrient requirements
Tomato (*Lycopersicon esculentum*)	Garlic and basil (to prevent mildew), spinach, beans, leeks, lettuce and peppers	Prefers good, moist soil; sunny to partial shade, high nutrient requirements
Jerusalem artichoke (*Helianthus tuberosus*)	Good around the edges of pumpkins and cucumbers; serves as a windbreak; good as a distraction plant near fruit trees for voles	Good, loose soil; frost hardy, pervasive (very competitive), high nutrient requirements
Wild asparagus (*Asparagus officinalis*)	Lettuce and other low-growing plant cover	Undemanding, hardy, sunny areas, frost resistant, wild vegetable
Skirret (*Sium sisarum*)	Onions, leeks, lettuce, carrots, parsnips and peas	Good, loose soil; sunny areas, hardy, frost resistant, vegetable
Onion (*Allium*)	Carrots, parsnips, skirret, lettuce, chicory, black salsify, radishes and beetroot	Good, loose soil; sunny areas, medium nutrient requirements

Medicinal and Culinary Plants

Plant	Characteristics and Requirements
Narrow-leaved inula (*Inula ensifolia*)	Good soil, sunny to partial shade, frost hardy, medicinal plant
Valerian (*Valeriana officinalis*)	Undemanding, but prefers moist soil, partial shade, frost hardy, medicinal plant
Balsam herb (*Tanacetum balsamita*)	Also thrives on poor soils, sunny to partial shade, frost hardy, medicinal and culinary plant
Sweet basil (*Ocimum basilicum*)	Good, loose soil; sunny areas, annual, culinary plant
Mugwort (*Artemisia vulgaris*)	Very undemanding, sunny areas, frost hardy, medicinal and culinary plant (good with pork)
Comfrey (*Symphytum officinale*)	Undemanding, but prefers good soils, sunny to partial shade, medicinal plant, liquid fertiliser
Mountain arnica (*Arnica montana*)	Marshy soil, undemanding, sunny areas, frost hardy, medicinal plant
Winter savory (*Satureja montana*)	Good, loose soil; sunny areas, hardy and frost resistant, culinary plant, keeps aphids away from beans
Borage (*Borago officinalis*)	Undemanding, but prefers good soils, annual, sunny to partial shade, good for improving heavy soils, medicinal and culinary plant
Watercress (*Nasturtium officinale*)	Moist to wet areas (on banks), partial shade, frost hardy, can be added to salads
Dill (*Anethum graveolens*)	Good, moist soil; culinary plant, sunny to partial shade, annual
Southernwood (*Artemisia abrotanum*)	Also thrives on poor soils, sunny areas, frost hardy, medicinal and culinary plant, tastes of lemon
Scented mayweed (*Matricaria chamomilla*)	Undemanding, annual, sunny areas, medicinal plant
Heath speedwell (*Veronica officinalis*)	Undemanding, poor soils, sunny to partial shade, frost hardy, medicinal plant
Marsh mallow (*Althaea officinalis*)	Good soil, sunny areas, does not get along with many other plants well, frost hardy, medicinal plant
Garden angelica (*Angelica archangelica*)	Good, deep, moist soil; partial shade, biennial, frost hardy, medicinal and culinary plant
Safflower (*Carthamus tinctorius*)	Also thrives on poor soils, annual, sunny areas, dye plant – yellow/orange flowers
Yellow chamomile (*Anthemis tinctoria*)	Undemanding, also thrives on poor and dry soils; sunny areas, frost hardy, dye plant – yellow flowers
Tarragon (*Artemisia dracunculus*)	Undemanding, but prefers good soils, sunny areas, sensitive to frost, culinary plant with a very intense flavour
Lady's mantle (*Alchemilla erythropoda*)	Very undemanding, sunny to partial shade, frost hardy, medicinal plant

Plant	Characteristics and Requirements
Yellow gentian (*Gentiana lutea*)	Undemanding, sunny areas, frost hardy, medicinal plant
Fennel (*Foeniculum vulgare*)	Good soils, sunny to partial shade, frost hardy, culinary and medicinal plant
Bergamot (*Monarda sp.*)	Also thrives on poor soils, sunny areas, frost hardy, medicinal plant, can be used to make tea
Dense-flowered mullein (*Verbascum densiflorum*)	Undemanding, sunny areas, biennial, frost hardy, medicinal plant
Motherwort (*Leonurus cardiaca*)	Undemanding and hardy, sunny to partial shade, frost resistant, medicinal plant
Houndstongue (*Cynoglossum officinale*)	Undemanding, prefers good, moist soils; biennial, sunny to partial shade, good decoy plant for voles
St. John's wort (*Hypericum perforatum*)	Also thrives on poor soils, sunny areas, frost hardy, medicinal plant
Sweet flag (*Acorus calamus*)	Marshy areas, on banks, sunny to partial shade, frost hardy, medicinal and culinary plant
Garden nasturtium (*Tropaeolum majus*)	Undemanding, sunny to partial shade, annual, can be added to salads, good decoy plant for aphids
Chervil (*Anthriscus cerefolium*)	Undemanding, annual, sunny areas, culinary and medicinal plant
Small-flowered willowherb (*Epilobium parviflorum*)	Also thrives on poor soils, sunny areas, frost hardy, medicinal plant
Garlic mustard (*Alliaria petiolata*)	Garlic flavour, undemanding, partial shade, frost hardy, medicinal and culinary plant
Coriander (*Coriandrum sativum*)	Good, moist soil; annual, sunny areas, culinary plant (fresh leaves and seeds)
Cornflower (*Centaurea cyanus*)	Undemanding, dry soils, annual, sunny areas, medicinal plant
Curly mint (*Mentha spicata var. crispa*)	Good soil, sunny to partial shade, frost hardy, medicinal plant, can be used to make tea
Horseradish (*Armoracia rusticana*)	Undemanding, sunny to partial shade, frost hardy, culinary plant
Lavender (*Lavandula angustifolia*)	Well-drained soils, sunny areas, frost hardy, discourages aphids, medicinal plant
Lovage (*Levisticum officinale*)	Prefers good soils, hardy, sunny to partial shade, frost hardy, culinary and medicinal plant
Meadowsweet (*Filipendula ulmaria*)	Good, moist to wet soils (on banks), partial shade, frost hardy, medicinal plant
Marjoram (*Origanum majorana*)	Good soils, sunny areas, medicinal and culinary plant

Plant	Characteristics and Requirements
Milk thistle (*Silybum marianum*)	Undemanding, sunny areas, biennial, frost hardy, medicinal plant
Masterwort (*Peucedanum ostruthium*)	Good, moist soils; partial shade, frost hardy, medicinal plant
Musk mallow (*Malva moschata*)	Also thrives on poor soils, sunny areas, frost hardy, medicinal plant
Feverfew (*Tanacetum parthenium*)	Undemanding, sunny areas, frost hardy, medicinal plant
Evening primrose (*Oenothera biennis*)	Undemanding, sunny areas, frost hardy, medicinal plant, flowers open at night – moths
Agrimony (*Agrimonia eupatoria*)	Well-drained soil, sunny to partial shade, frost hardy, medicinal plant
Oregano (*Origanum vulgare*)	Good, moist soil; sunny areas, frost hardy, culinary plant
Parsley (*Petroselinum crispum*)	Prefers good soil, partial shade, biennial, culinary plant
Peppermint (*Mentha piperta*)	Undemanding, but prefers good soil, sunny to partial shade, frost hardy, medicinal plant, can be used to make tea
Pennyroyal (*Mentha pulegium*)	Undemanding, but prefers good soil, sunny to partial shade, sensitive to frost, medicinal plant, can be used to make tea
Broad-leaved thyme (*Thymus pulegioides*)	Undemanding, sunny areas, frost hardy, medicinal and culinary plant (has a particularly intense flavour when growing on poor, dry soils)
Tansy (*Tanacetum vulgare*)	Also thrives on poor soils, sunny areas, frost hardy, can be used in liquid fertiliser
Pot marigold (*Calendula officinalis*)	Undemanding, but prefers good soil, annual, sunny areas, good plant to discourage nematodes in a mixed planting, medicinal plant
Purple coneflower (*Echinacea purpurea*)	Good, well-drained soil; sunny areas, frost hardy, medicinal plant (particularly in homoeopathy)
Sage (*Salvia officinalis*)	Undemanding, sunny areas, frost hardy, medicinal and culinary plant (has a particularly intense flavour when growing on poor, dry soils)
Garlic chives (*Allium ramosum*)	Wonderful flavour, undemanding, but prefers good soils, sunny to partial shade, frost hardy, culinary plant
Chives (*Allium schoenoprasum*)	Good soil, sunny to partial shade, frost hardy, culinary plant
Greater celandine (*Chelidonium majus*)	Undemanding, but prefers good soils, sunny to partial shade, frost hardy, medicinal plant
Black hollyhock (*Alcea rosea var. nigra*)	Beautiful hollyhock, nutrient-rich soils, sunny areas, medicinal plant, can be used to make tea
Soapwort (*Saponaria officinalis*)	Undemanding, hardy, sunny to partial shade, frost hardy, the roots can be boiled to make suds

Plant	Characteristics and Requirements
Sweet cicely (*Myrrhis odorata*)	Good, moist to wet soil; partial shade, frost hardy, culinary plant
Centaury (*Centaurium erythraea*)	Poor soils, sunny areas, biennial, frost hardy, medicinal plant
Thyme (*Thymus vulgaris*)	Undemanding, sunny areas, frost hardy, medicinal and culinary plant (has a particularly intense flavour when growing on poor, dry soils)
Sweet woodruff (*Galium odoratum*)	Good, moist soil; partial to full shade, frost hardy, culinary plant, can be made into sweet woodruff punch
Wild garlic (*Allium vineale*)	Very aromatic, sunny areas, frost hardy, culinary plant
Rue (*Ruta graveolens*)	Prefers loose soil, sunny areas, sensitive to frost, discourages insects, medicinal and culinary plant
Wormwood (*Artemisia absinthium*)	Also thrives on poor soils, sunny areas, does not get along with many other plants, frost hardy, medicinal plant
Arnica (*Arnica chamissonis ssp. foliosa*)	Undemanding, also on poor soils, sunny to partial shade, frost hardy, medicinal plant: effect similar to mountain arnica
Giant silver mullein (*Verbascum bombyciferum*)	Undemanding, sunny areas, biennial, frost hardy, medicinal plant
Herb hyssop (*Hyssopus officinalis*)	Also thrives on poor soils, sunny areas, frost hardy, medicinal and culinary plant
Betony (*Stachys officinalis*)	Undemanding, sunny to partial shade, frost hardy, medicinal plant
Lemon balm (*Melissa officinalis*)	Undemanding, but prefers good soils, sunny areas, frost hardy, medicinal plant, can be used to make tea

6 Projects

Scotland

The principles of permaculture work all over the world. In my first book *The Rebel Farmer* I have already described my projects in Brazil, Colombia and North America (Montana). Using strategies based on permaculture it becomes possible to practice agriculture successfully under difficult conditions (soil conditions, climate). The following account should give anyone that is interested in permaculture the courage to make their visions and plans a reality even if the locations they are working in are not 'favourable'. On my project in Scotland I could see the positive results that were possible on acid soil with no real work to maintain the area within a short period of time.

On my first survey of the plot we selected test areas at different altitudes (approximately 100m–350m above sea level) in which I could begin to experiment. The photo shows a discussion with Gernot Langes-Swarovski, Mag. Christian Koidl and my wife Veronika on site.

The permaculture project in the Scottish Highlands was in cooperation with the Langes-Swarovski family. The goal was to create a permaculture garden for their private use day to day. This shared project gave me the opportunity to try out my methods on the acid peat soil (the pH value was between four and five) of the Scottish heathland.

Finally in May 2004 the first raised bed was created with an excavator. I made sure to test the effect of the raised bed on the different soil types (peat to marshy soil) and at different altitudes in the area.

FACING PAGE – LEFT TO RIGHT, TOP TO BOTTOM:

In the photo an eroded section of land that we chose as a test area can be seen. It was fenced off to minimise grazing by deer. The vegetation in the area mostly consists of sedge (*Cyperaceae Eriophorum* and *Carex*) and heather (*Calluna vulgaris*), which cope well with the prevailing acid conditions.

In May 2003 we sowed these test areas with the first seed mixtures. They contained cereals (ancient cereals: emmer, einkorn and ancient Siberian wheat) with a catch crop of vegetables (radishes, lettuce etc.) and soil improving plants (various legumes). The intention was to find out if and how the selected plants would grow under the prevailing conditions of the area (constantly windy, acid and moist soil). In the photo: preparing to sow seeds with Peter Wemyss and a colleague.

After sowing, a layer of straw was spread loosely over the test area as mulch.

One year later, in May 2004, this area is completely transformed. The cereals as well as the catch crop have germinated and developed very well!

The result far exceeded my expectations. Even under these difficult conditions – not actually suited to plants at all – the catch crop of vegetables grew wonderfully!

Even gentian seeds (*Gentiana lutea* and *Gentiana punctata*) germinated in a sheltered spot and developed strong new plants after a year. As a result of this promising outcome we began to plan a test raised bed system.

ABOVE, FROM TOP:

Newly created raised bed on the heath. Bulky, cleared material (trees, tree stumps, branches from pine trees and spruces among others) and heather was introduced. The raised beds were positioned in the shape of a wavy line.

The new raised beds were covered with straw. The mulch layer did not only have the purpose of protecting the soil from the effects of the harsh weather, but also of protecting the seeds from being eaten by birds. As can be seen in the photo, the raised beds are in the immediate vicinity of a small road. This way the beds can be kept an eye on and are easily accessible. Also, the danger of the plants being eaten is lowered if the crop is near a road.

LEFT TO RIGHT, TOP TO BOTTOM:

My most recent visit to Scotland on 23rd July 2004: I harvested radishes with Mag. Christian Koidl (advisor to and project leader for the Langes-Swarovski family). The growth was lusher than expected; the ancient grain was already two metres tall! These yields were achieved on what was previously an eroded area, through the use of seed mixtures and mulch.

With my colleague Erich Auernig I also erected a storage cellar in the Scottish Highlands using my tried and tested system. The cellar is for storing produce, but can just as well be used as an open shelter.

Berkshire pigs (an old pig breed) and geese move into their accommodation in the new permaculture system.

Planning further ways to proceed with Gernot Langes-Swarovski. Since the most recent tests have brought astonishing results, we plan to introduce wild and cultivated fruit trees into the permaculture system. Over the course of time the depleted and eroded areas will develop into 'edible landscapes' with diverse flora and fauna.

SWAROVSKI

Herrn
Sepp Holzer
Krameterhof
Keuschnig 13
5591 Ramingstein

Wattens, 22. Juli 2004

Liebe Sepp,

zu Deinem Geburtstag gratuliere ich Dir sehr herzlich und wünsche Dir das Allerbeste, vor allem jedoch Gesundheit, Glück und Erfolg.

Ich habe es wiederum genossen, mit Dir in Schottland zu sein, meine Begeisterung über das Gedeihen des Projektes kann ich kaum in Worte fassen. Vielen, vielen Dank!

Es ist eine Freude, mit Dir zusammenzuarbeiten, und ich hoffe, daß wir noch viel gemeinsam umsetzen können.

Mit nochmaligen guten Wünschen

Dein

Christian

D. SWAROVSKI & CO.
A-6112 WATTENS/AUSTRIA · TEL. +43 (0) 52 24 / 500-0* · FAX +43 (0) 52 24 / 52 3 35 · BAHNSTATION FRITZENS/WATTENS, DVR.-NR. 0000302
INTERNET: http://www.swarovski.com

Dear Sepp,

My most heartfelt congratulations on your birthday and I wish you all the best. Most of all I wish you health, happiness and success.

I enjoyed seeing you again in Scotland, I cannot express in words the enjoyment the success of this project has given me. Thank you so much!

It is a joy to work with you and I hope that we can work on further projects together in the future.

With my best wishes once more,
Christian

Thailand

At the end of 2003 a request reached me from a doctor couple in Thailand. About 100 kilometres north of Bangkok they run an orphanage, which currently has 40 children in its care. The couple told me that they would like to increase the capacity of the orphanage to 100 children. The goal was for the orphanage to be able to grow enough food to make it self-sufficient. So in January 2004 I flew to Thailand with my own helpers and trainees to take a look at the situation in person and to support the family's plans.

When first surveying the area, the owners expressed their wish to incorporate the words 'LOVE ~ PEACE' into the project during the coming excavation work. As there is a flightpath going over the their land to Bangkok, the words should be large enough so that the message can still be seen from an airplane.

First I had to spend an evening taking a look at this proposal and its prospects. Then the idea occurred to me that we could make the letters out of raised beds and banks, then create ponds and ditches in the hollows inside the letters.

On the southern side of the words 'LOVE ~ PEACE' we plan to divert water into the first letter 'L' and divert it back out at the final letter 'E'. The letter 'E' will be shaped into the deepest pond. Within the individual ponds deep and shallow areas will be made to best fulfil the different requirements of the various types of fish, crayfish, crabs and mussels. Different depths of water mean that areas of the pond with different temperatures can be created. This way problems such as a lack of oxygen or the pond becoming overgrown can be minimised.

These changes to the shape of the land will not only increase the area of workable land, but also protect it from flooding. At present, it is flooded during the rainy season and cannot be worked. The aim is that the entire area of land will be protected from flooding by a dam, which will also minimise the amount

The area of land available for the permaculture project directly borders intensively farmed paddy fields.

PERMACULTURE PROJECT IN THAILAND

Cross section of a letter. The water level of the pond varies according to the time of year and the rainfall.

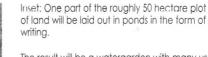

Inset: One part of the roughly 50 hectare plot of land will be laid out in ponds in the form of writing.

The result will be a watergarden with many uses. A lake with an island, on which an earth cellar will be built, will complete the plan. The system will be enclosed by a raised bed.

The edges of the letters are marked with poles, so that work with the excavator can begin.

The children help to plant fruit trees (mango, papaya and many others) in a trench for later use.

My colleague of many years Erich Auernig prepares a canopy for the trench.

of harmful substances coming in on the side bordering the heavily fertilised paddy field. These changes will also include the creation of different microclimates such as dry areas and 'mini rainforests'. The changes to the land should not only provide usable areas of land that can be farmed the whole year round, but also a recreational and experimental landscape as well as a garden for the children at the orphanage and the people at the nearby university hospital to relax in. The whole area can then be used as a pick-your-own area, a display garden and to keep livestock on. It can also serve as a sanctuary for birds and wild animals.

As high quality clay was available in such inexhaustible amounts, the possibility of utilising this raw material presented itself immediately. Mostly I thought of using it for building. Storage rooms and even houses could, according to my proposal, be built simply and very inexpensively. Using an excavator it would be possible to construct a building of this kind very quickly.

I considered restoring the water supply, which was polluted with many harmful substances, to its natural state to be particularly important. So my recommendation for the orphanage was to create a reed bed to purify its waste water and to regenerate and use the surface and ground water. For the water treatment itself I pictured treatment ponds with the appropriate plants, aeration and filtering. I also recommended the construction of a deep well. In order to save as much energy as possible, a pump system powered by water, wind power or electricity from photovoltaic cells could be used.

The commitment of the couple running this orphanage is really an example to us all. The question of the extent to which my ideas and suggestions have been implemented must be left unanswered at this point. As Thailand is too far away for me to be able to oversee the system on a regular basis or to participate in the ongoing work, it only remains to me at this point to wish for the success of the project that we began and worked on together with such joy.

TOP
Loam and clay is available on site in abundance and can be made good use of as a building material.

RIGHT
Water was available in large amounts, unfortunately it was heavily polluted from intensive farming. The photo shows a paddy field being farmed.

BELOW
Permaculture: a way for the future.

berⱡɑ: a Project for Lebenshilfe Ausseerland

A unique permaculture project is being undertaken in the area of Bad Aussee in Styria. The intention is to make it possible for disabled people to integrate in a special way with the help of permaculture. Using various techniques the project will make a contribution towards giving relief to the families of disabled people. I worked on the concept for this project with Lebenshilfe Ausseerland (under the leadership of Roland Kalß) in June 2003. I am happy that I can play a part in making permaculture accessible to disabled people and their families with my work on this community project. This is also why I agreed to give the Berta project my support.

The project includes the creation of gardens for growing many different vegetable and fruit varieties, various pleasant and peaceful areas for training the senses, an activity garden to teach mobility, a water garden and rocky area as a meeting place for visitors. There will also be an earth cellar and an earth shelter. The plans for this project can be found in the 'Landscape Design' chapter. I would now like to give a quick overview of the concept of the project.

Mediterranean Garden

In the Mediterranean garden indigenous as well as non-indigenous plants and trees will be grown in a series of different experiments. As a result of the unusual construction (terraced and protected from the elements, the intention of the

Children enthusiastically helping to plant and make bridges in the Mediterranean garden.

design is to store heat) the average annual temperature in this area should be higher and therefore make it possible to grow grapes, figs, kiwifruit and other fruits that require sunny conditions.

Water Garden

This area should bring our disabled visitors closer to the medium of water with all its plant and animal inhabitants. Direct contact should be the focus. This can be achieved by creating a ford that can also be crossed in a wheelchair, and therefore also makes it possible to plant from a wheelchair. As well as the propagation of aquatic plants, the breeding of fish is also planned in the water garden.

Sensory Garden

The specially selected plants and building materials in this garden will bring pleasure to the visitors' senses. The garden will be planted with heavily scented and aromatic flowers and herbs, as well as delicious berries and fruit. This will delight the visitors' senses of smell and taste. Brightly coloured flowers and eye-catchers will lead people's eyes through the labyrinth. The visitors' sense of touch will be inspired through the selection of different natural materials (stone, wood and water).

The sensory garden is currently under construction and has a view of the Mediterranean garden. The rocks in the garden need to be placed more irregularly and the raised beds still need to be finished.

The structure of the sensory garden: the remaining work mostly needs to be carried out by hand.

Activity Garden

In this area the visitors' mobility will be stimulated. Balance, coordination and fine motor function will be taught in equal measure using natural aids (different rock formations and wooden structures).

Rest Area and Rocky Area

This area should attend to the spiritual lives of the visitors and encourage them to relax through the careful selection of naturally available positive vibrations. While the rest area will be reserved for physical well-being (eating outdoors, picnics) among other things, the rocky area will become a cultural meeting point. Music and poetry will find their place here.

Earth Cellar and Shelter

The earth cellar is for storing produce from the permaculture area. The positioning and equipment of this building makes it ideally suited to storing fruit and vegetables, these factors also make it possible to offer fresh produce on site until late in the winter. The shelter should also be suitable as an open shelter for livestock if needed (possibly pigs).

Our disabled visitors will be included in the project from the outset. It is also planned that they will help with the project in practical terms, such as by planting different areas and creating environments where they can be happy and comfortable. The entire area will be planned so that as many aspects of planting and harvesting as possible can be achieved from a wheelchair. An important part of the integration process could be the marketing of produce grown on site to the local population at a local weekly market or at a farmers' market.

On 22nd April 2004 the groundbreaking work for the Berta project took place. I now look forward to seeing the fruition of this project.

Concluding Thoughts

Working towards a natural life and natural agriculture is difficult in a time when people in agriculture, science and politics only have their eyes on 'progress', whilst showing no consideration for nature. 'Grow or give way' is the motto of the modern age. In our competitive economy there is little space left for natural thinking.

Livestock are kept in increasingly cramped conditions, feeding is automated and controlled by computers. So contact with the livestock is lost and the animals are seen as a commodity rather than as living creatures. People simply refer to it as 'meat production'. You can only expect healthy produce from a healthy animal. Animal suffering is passed on to people. My observations confirm this again and again. Treating our environment and fellow creatures with respect is the only proper way.

One of my central ideas is: 'Try putting yourself in the position of your fellow creatures, whether they are plants or animals, and you will quickly find out whether the environment that you intend for them is right or not. If you observe a plant or animal closely, you will quickly see if it is happy. However, if you would not want to live in that environment as a plant or animal, then change the living conditions there quickly! Only animals that live happy lives will work for you day and night and you will be the biggest winner as the owner of a healthy plant and animal kingdom.'

There is still so much for me to say about my experiences working with plants and animals. Unfortunately, a book cannot contain it all. When we first met nine years ago, my friend Professor Bernd Lötsch – in many respects also my role model – asked me to document all of my practical experiences. I will try, as far as I can, to keep the promise I made back then by recording and passing on my experiences and observations.

I hope that this book contributes to both nature and the world being treated with more respect. Nature is perfect in all of its creation, only we humans make mistakes.

The Authors

Josef ("Sepp") Holzer was born in the province of Salzburg, Austria. He is a farmer, author, and an international consultant for natural agriculture. He took over his parents' mountain farm business in 1962 and pioneered the use of ecological farming techniques, or permaculture, at high altitudes (roughly 5,000 feet above sea level) after being unsuccessful with regular farming methods. He has been called the "rebel farmer" because he won't back down to conventional agricultural systems—despite being fined and threatened with prison for practices such as not pruning his fruit trees. Holzer conducts permaculture seminars at his farm and worldwide, has written several books, and is the subject of the film *The Agricultural Rebel*. He works nationally as a permaculture activist in the established agricultural industry, and works internationally as an adviser for ecological agriculture.

COURTESY OF PERMANENT PUBLICATIONS

The Co-Authors

Mag. Claudia Holzer received her secondary school education in Tamsweg and went on to study biology (specialising in zoology) in Graz. For her dissertation she studied the diversity of species of insect in the raised bed systems on the Krameterhof. From 2002 she has been working as an independent biologist in the areas of ecological education and permaculture.

Josef Andreas Holzer attended the School of Forestry in Bruck an der Mur. After completing his final diploma exam, he began studying ecology and bio-diversity in Graz in 2002. In addition to his studies he also works intensively with permaculture principles.

Index

Positive Inspiration for Making a Better World

Permaculture Magazine helps you live a more natural, healthy and environmentally friendly life.

Permaculture Magazine offers tried and tested ways of creating flexible, low cost approaches to sustainable living. It can help you to:

- **Make informed ethical choices**
- **Grow and source organic food**
- **Put more into your local community**
- **Build energy efficiency into your home**
- **Find courses, contacts and opportunities**
- **Live in harmony with people and the planet**

Permaculture Magazine is published quarterly for enquiring minds and original thinkers everywhere. Each issue gives you practical, thought provoking articles written by leading experts as well as fantastic ecofriendly tips from readers!

permaculture, ecovillages, ecobuilding, organic gardening, sustainable agriculture, agroforestry, appropriate technology, downshifting, community development, human-scale economy ... and much more!

Permaculture Magazine gives you access to a unique network of people and introduces you to pioneering projects in Britain and around the world. Subscribe today and start enriching your life without overburdening the planet!

Every issue of *Permaculture Magazine* brings you the best ideas, advice and inspiration from people who are working towards a more sustainable world.

Permanent Publications
The Sustainability Centre, East Meon, Hampshire GU32 1HR, UK
Tel: 01730 823 311 Fax: 01730 823 322
(Overseas: int code +44-1730)
Email: subscriptions@permaculture.co.uk
Web: www.permaculture.co.uk

THE QUEEN'S AWARDS
FOR ENTERPRISE

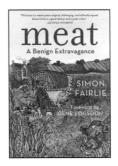